CENTRE FOR CO-OPERATION WITH THE ECONOMIES IN TRANSITION

SCIENCE, TECHNOLOGY AND INNOVATION POLICIES

FEDERATION OF RUSSIA

VOLUME I

EVALUATION REPORT

ORGANISATION FOR ECONOMIC CO-OPERATION AND DEVELOPMENT

ORGANISATION FOR ECONOMIC CO-OPERATION AND DEVELOPMENT

Pursuant to Article 1 of the Convention signed in Paris on 14th December 1960, and which came into force on 30th September 1961, the Organisation for Economic Co-operation and Development (OECD) shall promote policies designed:

- to achieve the highest sustainable economic growth and employment and a rising standard of living in Member countries, while maintaining financial stability, and thus to contribute to the development of the world economy;
- to contribute to sound economic expansion in Member as well as non-member countries in the process of economic development; and
- to contribute to the expansion of world trade on a multilateral, non-discriminatory basis in accordance with international obligations.

The original Member countries of the OECD are Austria, Belgium, Canada, Denmark, France, Germany, Greece, Iceland, Ireland, Italy, Luxembourg, the Netherlands, Norway, Portugal, Spain, Sweden, Switzerland, Turkey, the United Kingdom and the United States. The following countries became Members subsequently through accession at the dates indicated hereafter: Japan (28th April 1964), Finland (28th January 1969), Australia (7th June 1971) and New Zealand (29th May 1973). The Commission of the European Communities takes part in the work of the OECD (Article 13 of the OECD Convention).

The Czech Republic, Hungary, Poland and the Slovak Republic participate in the OECD Programme "Partners in Transition", which is managed by the Centre for Co-operation with the Economies in Transition.

Publié en français sous le titre :
POLITIQUES DE LA SCIENCE,
DE LA TECHNOLOGIE ET DE L'INNOVATION
FÉDÉRATION DE RUSSIE
Rapport d'évaluation

© OECD 1994
Applications for permission to reproduce or translate all or part of this
publication should be made to:
Head of Publications Service, OECD
2, rue André-Pascal, 75775 PARIS CEDEX 16, France

Foreword

At the request of the Minister for Science and Technological Policy of the Federation of Russia, the Centre for Co-operation with the Economies in Transition (CCET) included a review of science, technology and innovation policies in the Federation of Russia in its programme of assistance to Russia. This review aims to assist the Russian authorities to define a new policy in these areas. This activity was implemented by the OECD Directorate for Science, Technology and Industry, which drew on expertise in the OECD Secretariat, as well as the expertise in Member countries.

The review follows the model adopted for the country studies carried out by the OECD. The objectives and areas to be covered were defined at a meeting of representatives of the Russian Government and of Delegates of Member countries, held at the OECD Headquarters in December 1992. A "Background Report" has been prepared by a group of Russian experts under the supervision of A.G. Fonotov, First Deputy Minister of Science and Technology Policy; it describes the Russian science and technology system, recent developments, and the Government's policies. This constitutes Volume II of this publication. The "Evaluation Report", which constitutes Volume I of this publication, has been prepared by a group of experts from OECD countries led by Professor J.-L. Lions, Collège de France, President of the International Mathematical Union. J.-E. Aubert of the OECD's Directorate for Science, Technology and Industry co-ordinated the work on the study.

The Evaluation Report discusses general, strategic and institutional aspects of government policies; it identifies main problems and formulates suggestions concerning policy actions that might help to resolve them. It does not attempt to evaluate Russian competence in various S&T fields, nor does it address important safety issues, such as the safety of nuclear power stations, which have been studied extensively by other international organisations. Instead, it offers information and policy analyses aimed at stimulating Russian policy makers to consider fresh approaches to policy problems as well as promoting Russia's dialogue with the international community.

The final report was discussed with the Russian authorities and experts at a meeting held in Moscow on 21-22 September 1993. A brief resumé of the discussion is presented as an annex to Volume I.

This report is published on the responsibility of the Secretary-General of the OECD.

Salvatore Zecchini
OECD Assistant Secretary-General
Director of CCET

Preface

Russian science, one of the greatest creations and the most valuable possessions of civilisation, is now going through a very difficult period.

For many decades the organisation of science, its structure and the psychology of the scientific community had to adjust to the conditions of the administrative system and thus acquired its features. The reform of science was and still is as necessary as the reform of society as a whole.

But the real process of reform, with its difficulties and failures, is extremely painful for the Russian scientific community. There is a danger of destroying many outstanding scientific schools, creative teams and important areas of scientific research of both national and world value. That is why preserving Russian science is now not only a national goal but also a concern for the world community.

Regarding implementation of scientific and innovation policies as one of its priorities, the Government of the Russian Federation accords great importance to the evaluation of its efforts by the world scientific community and authoritative international organisations.

The Organisation for Economic Co-operation and Development, under whose aegis the present report has been drafted, has much experience and is well known for the very high level of experts involved in its activities. The study devoted to science, technology and innovation policies of Russia is no exception and contains constructive recommendations that merit thorough study and use. We are sure that, through honest and open joint discussions of the problems of Russian science, effective solutions will be found.

B. Saltykov
Minister for Science and
Technological Policy
Federation of Russia

Table of contents

Foreword 3

Preface 5

Introduction 11

Chapter I

Major issues

1. A formidable policy challenge 15
2. An oversized, ill-adapted system in rapid deterioration 16
3. An unstable context 19
4. A dramatic opening to the outside world 21
5. Cultural and social obstacles to transition 22
6. Concluding remarks 24

Chapter II

The policy framework

1. Guiding institutions: from the old system to the new 25
2. New actors 30
3. The science budget 32
4. Priorities 35
5. International co-operation 36
6. Concluding remarks 41

Chapter III
The science base

1. A profound crisis.................................... 43
2. The Academy of Sciences........................ 45
3. University research 47
4. The branch institutes 48
5. Evaluating researchers, projects, institutes.... 50
6. Integrating Russia into the international scientific community 51
7. Brain drain and brain waste 54
8. Support to young graduates..................... 57
9. The social sciences and the humanities....... 58
10. Concluding comments 60

Chapter IV
The innovation climate

1. An embryonic market-based innovation system 61
2. The intellectual property framework 64
3. Support for innovators and entrepreneurs 66
4. The involvement of foreign industry........................... 69
5. Normalisation of technology: standards, certification, quality control 71
6. Concluding remarks... 72

Chapter V
From military to civil applications

1. Military S&T... 75
2. The problem of the closed cities 77
3. Space and aviation: the shift from a military to a civil focus...... 78
4. Developing civil S&T 80
5. Telecommunications: a key investment 83
6. Concluding comments 84

Chapter VI

Principal recommendations

1.	Reorganising the scientific and technological policy institutions....	88
2.	Adjusting resources invested in R&D to economic capacities	89
3.	Promoting and protecting innovation............................	91
4.	Developing and modernising civil technology...................	93
5.	Internationalising Russian science and technology	93
6.	Improving statistics and accounting practices	96

Annex 1

Tables

1.	R&D expenditure in Russia in 1990 and 1991	99
2.	Distribution of R&D personnel by USSR republics: 1980, 1985, 1990...	99
3.	Geographic distribution of Russia's R&D expenditures and personnel by economic regions, 1991	100
4.	Researchers employed by different types of performing organisations, 1990-91...................................	101
5.	Financing of R&D projects by type of activity and performer, 1990 and 1991	102
6.	Approximate number of institutes in the Russian R&D system	102
7.	Average monthly salaries in the sector "Science and scientific services"	102
8.	Payments for and receipts from trade in licences	103
9.	Estimated budgets of bilateral collaboration programmes in science and technology of selected OECD countries with Russia	103

Annex 2

Summary of the discussion of the evaluation report

Summary of the discussion of the evaluation report, Moscow, 21-22 September 1993	105
List of participants	113

List of boxes

A. How much can Russia afford to invest in R&D? 17
B. What is the extent of scientific and technical co-operation between OECD countries and Russia? 37
C. How extensive is the emigration of Russian scientists and technicians? 55

Introduction*

The Russian Federation is currently engaged in a vast transformation, a transformation in which science and technology are one of its greatest assets. The Soviet Union's scientific capacity, some 70 per cent of which was located in Russia, mobilised human resources on a scale which, at least by its dimensions, equalled or exceeded those of the major OECD countries. Even if that capacity must be considerably reduced, and reoriented from military to civil applications, it is a resource of immense importance and one on which the restructuring of the country must be founded. Indeed, next to the natural resources with which Russia is abundantly endowed, it may well be its principal treasure.

Saving, and taking the best advantage of, this scientific and technical capacity depends, however, on the establishment of an efficient economy and the realisation of a myriad of indispensable reforms in the legal and institutional framework, the banking and fiscal system, privatisation and financial stabilisation. Achieving these goals obviously depends on the existence of a pluralist democracy that is sufficiently politically stable to guarantee that actions, once undertaken, can be carried through. Diplomatic efforts in favour of peace must also be pursued, given the new responsibilities that the Russian Federation has assumed within the international community and in light of its special position among the New Independent States. More generally, Russia's decisions about the

---

* The principal examiners contributing to the report were: H. Balzer, Director of the Russian Area Studies Program, Georgetown University; J. Cooper, Director of the Centre for Russian and East European Studies, Professor, Birmingham University; and I. Linnakko, Managing Director, Oy Sitrans, Helsinki.
In the light of observations made at the review meeting in Moscow, minor modifications have been introduced into the original text of the report.

size and nature of its future S&T efforts would benefit from a clearer view of the role that it wishes to play on the world stage.

All is certainly far from perfect in the "western" world. For example, in a highly competitive globalised economy that makes ever greater use of automated technology, the constant search for optimisation has created growing unemployment which is proving difficult to curb. However, the market economy – not unrestrained capitalism, but a process whereby economic actors and individuals freely choose to allocate resources within a framework of regulations established by the State – seems to be the only alternative to the centralised and planned economy, of which the Soviet Union and the socialist countries have had the painful experience.

As it embarks upon this economic and political change, Russia must find its own way, taking into account its unique history and its socio-cultural traditions. There are many systems founded on democratic pluralism and the market economy, and there are vivid contrasts to be found in the OECD countries as one passes from North America to Japan, through Europe's Anglo-Saxon, Latin, Germanic and Scandinavian variants.

For Russia, finding its own way means, for example, discovering an appropriate balance between community and individual interests as they manifest themselves in entrepreneurship. It means instituting a carefully thought out and controlled process of privatisation, without falling into the ideological excesses of economic liberalism. It means putting to good use the tradition of a strong State – even if the State is now considerably weakened – able to make the administrative, legal, and judicial rules necessary to an efficient market economy. It means using existing structures and collective tendencies to cushion the social costs of the economic transformation and to satisfy the basic needs of the population. It means inventing a new administrative framework for incorporating those regions – long steeped in ancestral traditions and later closed off by the former regime – which now enjoy great autonomy. It means, finally, judiciously opening its frontiers to the world economy in order to benefit from its material and intellectual wealth, as Russia, imbued with European culture, did once before, several centuries ago.

Unless there is a profound and sustained process of reform, it is to be feared that the economy will continue to break down, thereby giving rise to more extensive and even more uncontrollable black market activities. The disintegration of society will continue. The risks of major ecological accidents – of which Chernobyl was a foretaste – will increase and imperil all humanity.

In the transition towards a new system, the principal sources of resistance may be the corporatist traditions and the patterns of allegiance which have long structured economic and social life and which are now directed towards maintaining acquired advantages or appropriating new sources of wealth in a context of crisis and penury. This behaviour formerly found its support in the Communist Party. It has reappeared in a more informal and disorderly fashion, but it remains strong and can be found everywhere. It is present in the scientific and technical community, itself very active in current political debate, although it carries less weight in a society currently more concerned with the short term.

While economically inefficient, Soviet science and technology had its organisational logic and produced outstanding achievements in certain fields. Today, ensuring the smooth introduction of another logic represents an extraordinarily difficult challenge. The scale of the restructuring that must be accomplished should not be underestimated. The research system should perhaps be reduced by half, compared to its size in the late 1980s. As for the linear, hierarchical, and compartmentalised "innovation system" inherited from the Soviet Union, it is completely unsuited to a market economy. The more appropriate model now being put in place is still in embryonic form, and it is likely to be at least ten years before it makes a significant contribution to strengthening the economy as a whole.

In the support that the international community is endeavouring to give Russia – an extremely important strategic issue for many countries – scientific and technical co-operation plays a significant role. In fact, public and private transfers from abroad are an essential source of funding for many institutes, scientists and entrepreneurs, and they can facilitate the adaptation of the research and innovation system. In return, access to Russian (and, more generally, ex-Soviet) science and technology is an event of the greatest importance for world science and technology and could have far from negligible benefits for the countries that engage in co-operation.

*

* *

In this Evaluation Report, the problems have been analysed as lucidly as possible and realistic policy proposals put forward, taking account of the unique characteristics of Russia and of the very particular situation in which it currently finds itself. Every attempt has been made to avoid proposing measures that may have proved their worth in OECD countries but do not seem transposable to Russian conditions. Also, the report has been focused on general principles, without entering into the details of the actions proposed.

Chapter I provides an overview of the problems that affect science and technology in the Russian Federation. Chapter II analyses the institutional and financial framework in which government policy is developed. Chapter III focuses on the reorganisation of research structures, drawing attention to the worrisome question of the fate of the scientific work force. Chapter IV examines the climate for innovation in the difficult transition to a market economy. Chapter V considers the gradual shift from military to civil applications of science and technology, as the basis for development. Chapter VI sets out the Report's principal recommendations.

All statistical data presented in this Evaluation Report are extracted from the Background Report, with the exception of the sources noted below.[1] Annex 1 contains a series of tables that outline major quantitative features of the Russian S&T system.

Note

1. These additional sources are:
   - J. Cooper, *The Conversion of the Former Soviet Defence Industry*, London, Chatham House, 1993 (This source is used in Chapter II, Section 3 and Chapter V, Sections 1 and 2 of the present report).
   - OECD Secretariat enquiries (compiled by S. Prerost) on S&T international co-operation (Chapter II, Section 5 and Box B).
   - A report prepared for the EC (by L. Vandenbrouck and L. Calot) "The Space Activities of the CIS", January 1993 (Chapter V, Section 3).

Chapter I

Major issues

1. A formidable policy challenge

In restructuring its S&T system, the post-Communist Russian Federation faces formidable problems. These problems are more severe than those encountered by the countries of central and eastern Europe. It is not simply a question of scale, although the vast size and extraordinary regional diversity of the Russian Federation are without doubt complicating factors. Russia has inherited a set of S&T institutions and practices that were profoundly shaped, over more than 70 years, by the Soviet experience. While the inheritance of central and east European countries was similar, the Soviet-type structures had been less profoundly institutionalised and had often been set on foundations more favourable to transformation. In Russia, the state-dominated administrative-command economy – with its emphasis on the military sector and heavy industry and its characteristic extensive growth pattern – generated an extremely large and highly distorted S&T base. In setting S&T priorities, the Communist Party valued the instrumental contribution of S&T to the realisation of its goals, but, at the same time, sought to contain the intellectual endeavours of the research community within strict ideological limits. In particular, international links were subject to central control, and many scientists, especially those working in strategically sensitive fields, were isolated from their colleagues abroad. Scientists themselves had very little control over the development of S&T. Despite these limitations, Russian scientists and engineers have S&T achievements of the first magnitude to their credit. In making the transition to a democratic order founded on a market economy, Russia needs to transform science and technology in such a way that the best will be retained to contribute to the country's economic, social and cultural revival.

The processes of transition threaten the very survival of the Russian S&T system. In 1992, GDP declined by one-fifth, price increases led to hyperinflation, and state budget expenditure came under severe pressure. State funding of S&T was reduced dramatically, and other sources of finance for S&T virtually disappeared. The traditional plans and state orders for S&T have dried up. Faced with acute problems of short-term survival, enterprises have no interest in innovation and possess no financial resources for financing R&D. Under these conditions, there is little scope for market-led S&T activity. To make matters worse, the breakup of the USSR has deprived many Russian S&T organisations of their traditional links with organisations now located in other independent states of the former Union. These circumstances pose a formidable policy challenge: how to restructure S&T to make it compatible with a market economy while surviving deep economic recession.

2. An oversized, ill-adapted system in rapid deterioration

By comparative international standards – in terms of employment, number of establishments and the resource commitment as a share of GNP – the S&T base of the former Soviet Union was extremely large. The systematic distortions of the price system tended to understate the cost of high technology activities, especially those concerned with the military/space sector. Furthermore, the statistical methods used to measure S&T activity were seriously deficient and incompatible with international practice. As a result, the actual costs of maintaining the vast R&D network were known neither to the country's leadership nor to its citizens.

The expansion of S&T institutions tended to be driven not by economic considerations but by the value attached to technological prestige and by the bureaucratic interests of state administrative hierarchies. Once established, R&D organisations grew inexorably, following the pattern of extensive growth typical of the whole economy. There is no doubt that, in relation to the scale of the economy and its real level of development, Russia now has an excessively large S&T sector (see Box A). The Government must determine what level of S&T is appropriate to the Russian Federation as a new democratic state with a market economy in a changed international environment. To do this, they urgently need more accurate and internationally comparable statistics. Privatisation and the

Box A

How much can Russia afford to invest in R&D?

The Soviet Union, like the other countries of eastern Europe, considered R&D important. According to published data (see the Background Report), some 3.5 per cent of GNP was allocated to R&D towards the end of the 1980s, a proportion markedly higher than that of the OECD countries, even among those that invest the most. However, when compared to OECD norms, these figures overestimate the real expenditure on R&D. First, they are based on a broader definition of R&D. Second, GNP according to the Soviet statistics is considerably lower than the OECD equivalent. Once the figures are corrected to correspond to OECD norms, R&D investment was approximately 2.1 per cent of GDP in 1990 and fell to 1.4 per cent in 1991. This places Russia at the level of the OECD country average.

One may ask to what extent the R&D effort can or should be reduced in an economy that is completely integrated into the world economy and operates according to its costing standards. It is therefore important to situate Russia in terms of per capita income and compare its situation with that of countries at comparable levels of wealth. The calculation has certain limits if the real exchange rate is used, because account is not taken of important differences in purchasing power. It is better to use data adjusted for purchasing power parity (ppp). The International Monetary Fund has recently published figures for GDP which incorporate their ppp for countries around the world*. These calculations obviously increase the relative weight of developing countries and countries in transition. These new estimates show that the United States has 22.5 per cent of world wealth (against 27 per cent in the earlier type of calculation), the former Soviet Union 8.3 per cent, Japan, 7.5 per cent, China 6 per cent, Germany 4.5 per cent.

These data using ppp are only moderately useful for proposing gross amounts for investment in R&D. However, they do allow reasonable estimates of the number of researchers and technicians that Russia can afford to keep. The United States employed in the late 1980s some 1 million research scientists and engineers (RSE) in full-time equivalent (FTE) for an economy which, in real terms, seemed three times bigger than Russia's. Japan, whose science system is more labour-intensive, employed some 450 000 RSE for an economy which seemed equivalent, in real terms, to Russia's. These data give some benchmarks for the size that seems reasonable for Russia. The latter still employed 950 000 RSE in head count in 1991 – the corresponding number in FTE would be somewhat smaller. The reduction that has already occurred concerns possibly 200 000 to 300 000 persons, so that large cuts are still to come.

In conclusion, in financial terms, measured as a percentage of GNP or of the State budget, the R&D effort no longer appears excessive. However, the number of personnel employed, and the infrastructure, still appear much too large.

World Economic Outlook, Annex IV, IMF, Washington D.C., May 1993.

development of market institutions will also make it necessary to modify procedures for obtaining relevant statistical data.

Russian S&T is heavily biased towards the requirements of the military sector. Not only is the defence industry the major R&D performer, but substantial military research is undertaken by the institutes of the Academy of Sciences and by higher education establishments, especially the country's leading technical institutes. In recent years, approximately two-thirds of all state budget funding for S&T has been for military purposes. The budget of the nuclear industry alone has been comparable to that of the Academy of Sciences. With reduced levels of military expenditure, it becomes urgent to civilianise at least part of defence R&D.

Civilianising science will be aided by a new, more narrowly focused regime for the protection of military secrets. There is no doubt that the extraordinarily elaborate systems for security and ideological control maintained by the former KGB had a negative impact on the development of Soviet S&T. Partial reforms have been undertaken, but more action will be needed to secure conditions for the free flow of scientific information. Temptations to maintain past practices in the name of "commercial secrecy" should be resisted.

The former Soviet Union's innovation system was characterised by a top-down approach and considerable fragmentation. Scientific research, design activities, experimental development, and industrial production were considered as separate activities and were carried out in isolation from one another. Moreover, in the different sectors, tasks were distributed monopolistically to units located throughout the territory. The great majority of R&D structures were situated in Russia, and development and production were located in other republics. Material resources were allocated by administrative, non-market means, prices were fixed centrally and did not reflect resource scarcities, customer power was weak at best (*e.g.* in the defence sector) or non-existent, and there was no competition. Enterprises and R&D organisations were subject to soft budget constraints: failure to innovate never resulted in bankruptcy or closure. Enterprise managers were rewarded, above all, for maximising current output, and innovation was likely to reduce bonuses. In this supply-driven economic system, the central economic agencies and ministries had to resort to administrative pressure or one-off financial inducements to obtain innovation. R&D organisations and enterprises created and adopted new technology according to plan, often with little regard to economic costs and benefits.

Experience in other countries provides plenty of evidence that the existence of a market economy may not be a sufficient condition for rapid technological change. Government action is often required to create a climate for successful innovation. However, as Soviet experience amply testifies, problems of innovation are much more acute in the absence of capital, labour, and product markets. The former Soviet planned economy lacked an inherent mechanism for generating technological innovation. This non-market system of R&D and innovation will not be easily transformed. Fundamental changes in institutions, behaviour and attitudes will be necessary, but there may be an understandable temptation to maintain a substantial measure of state involvement of the traditional type. The State will retain a significant role, but will have to work in new ways to foster a climate promoting success in innovation.

The Russian Federation's well-educated population and considerable intellectual and cultural talent offer hope that the present acute transitional problems can be successfully overcome. At the same time, however, the processes of transition pose a threat to the country's educational base. Economic decline and state budget stringency are undermining financial support to the educational system at the very time when it is striving to adapt to new demands. Given the acute problems of the S&T system, the abrupt cut-back in the military-space sector, and a general decline in the prestige of science in Russian society, many young people are turning away from the natural sciences and engineering and towards disciplines or training in the new skills required by the emerging market economy. If the transition period is protracted, the future S&T capability of the country could be threatened.

3. An unstable context

While the Russian Government understands the general strategic requirements for restructuring S&T, as its policy orientation demonstrates, the overall transitional environment is not favourable to effective implementation of practical solutions. The political context is extremely unstable, and the division of power among the President, the Government and the parliamentary bodies – the Congress of People's Deputies and the Supreme Soviet – is unclear. Some policy issues can be resolved by Presidential decree, while others require parliamentary approval, which usually involves delays. Government ministries and state committees change frequently, and the lack of stability in organisations and high-level personnel complicates policy making. For example, in early 1993, the Ministry of Science, Higher Education and Technological Policy was deprived of its responsibility for higher education policy, and its minister was demoted from his previous position as deputy prime minister.

When laws are passed, the structures of executive authority are often insufficient to secure their implementation throughout Russia. In the past, the Communist Party provided a nation-wide structure for overseeing the enforcement of centrally determined policies. The formal breakup of the Party's networks, as well as the weakening of the central government – now unable to collect taxes as it did in the past – has led to a *de facto* decentralisation to regions and enterprises. The problem is exacerbated by the absence of agreed constitutional arrangements for the Russian Federation. The restructuring of S&T would be greatly facilitated by progress in establishing a stable and effective legal framework.

In general, institution building remains a central issue for the Russian transition. It is difficult to reshape S&T to make it compatible with a market economy when so many of the basic institutions of the latter – the banking sector, the labour and capital markets, and the methods of enterprise accounting and auditing – remain absent or embryonic. The weak institutional infrastructure reduces the effectiveness of market instruments for promoting innovation or for realising industrial policy. It will take time to establish an appropriate institutional environment and stable "rules of the game", and technical assistance from abroad clearly has a large role to play.

Inflation, which has now reached the astonishing rate of 2 000 per cent a year, adds significantly to the instability of the environment. It has a marked effect on scientific and technical workers, who continue to receive their salary from the State. It makes any kind of planning over periods of more than a few weeks impossible and presents many other disadvantages. It is true that the instability and intensified monetary liquidity have been profitable to a new breed of economic actors who have sprung up in the shadow of the large bureaucracies, including small entrepreneurs well connected to the financial networks. Certain observers believe that Russia will long be burdened with high and uncontrollable rates of inflation and foresee scenarios not unlike those that have plagued Latin American countries since the middle of the 1970s, in which governments alternately print money to keep up with inflation and place severe controls on prices and salaries.

4. A dramatic opening to the outside world

Just as the former Soviet economy was relatively closed to the outside world, so also was its science and technology isolated to a considerable degree from the international community. The Communist regime largely self-imposed this relative isolation, but COCOM and other Western defensive constraints reinforced it. During the Gorbachev period, isolation began to weaken and possibilities arose for more extensive research contacts and foreign business involvement in the economy. Since the collapse of the USSR, Russia's borders have opened rapidly, with direct and at times dramatic consequences – not always negative – for the R&D sector.

Isolated for decades from world market forces, Russian industry is now subject to international competition, and serious weaknesses in the domestic S&T capability have been revealed. In no sector has this been more evident than in microelectronics and computing. The ready availability of foreign computers has had a devastating impact on the relatively backward Russian industry. As a result, Russia is faced with serious issues of industrial policy, including policy for the protection of domestic high-technology sectors now threatened with rapid erosion.

The change has also presented citizens with new possibilities for emigration or temporary periods of work abroad. A substantial "brain drain" was feared, but the number of permanent emigrés does not yet appear to be dramatic. There is understandable concern in Russia about a damaging loss of scientific talent, but the present situation is not without its positive side. Many Russian scientists, especially in the younger generation, are becoming quickly integrated into the international scientific community. The changed situation has also confirmed that Russia does indeed possess considerable scientific talent, and this is a source of optimism for the future.

The opening up of Russian S&T has offered new opportunities for international collaboration. While there is concern in Russia that some foreign companies are seeking to exploit the country's S&T capabilities by paying Russian researchers low salaries or low prices – by international standards – for advanced technologies, the new situation clearly offers possibilities for forms of collaboration that will both ease the problems of transition and begin the process of integrating Russia into the global economy.

5. Cultural and social obstacles to transition

It is not surprising that the Soviet system has left a residue of cultural and social attitudes that may complicate transition to new S&T structures. In some cases, in fact, Soviet ideology and practices kept alive, or even reinforced, more traditional Russian cultural patterns.

In Soviet ideology, science and technology were always given considerable emphasis: the "scientific and technical revolution", after all, was to secure the final triumph of Communism on a global scale. Associated with this ideological conviction were strongly expressed scientistic and technocratic modes of thought which outlive the Soviet regime. There is thus a tendency to view economic growth as being, above all, a matter of developing and "introducing" new technology, to the neglect of other economic, social, and cultural factors. Technological avant-gardism – the view that only the best is acceptable, regardless of the economic rationality of adopting frontier solutions – remains widespread. In the Soviet system, a distorted price structure, which concealed true opportunity costs, served to reinforce these attitudes.

The abrupt collapse of the USSR has meant that Russia now has to come to terms with a loss of "empire". The USSR was generally regarded, at home and abroad, as a superpower, even if this status owed more to military might than to economic strength and vitality. It is understandable that many in Russia wish to see their country retain at least "great power" status. This is often interpreted to mean that Russia must remain a strong military power with a defence capability at the forefront of world technology. This conviction reinforces the general predisposition towards technological avant-gardism.

Because the military technology developed in the former Soviet Union was generally competitive with that of leading Western countries and relatively more advanced than Soviet civil technology, much of the Russian defence sector is genuinely convinced that its R&D is at the leading edge of world technology. It is widely believed that maintaining and further developing this military R&D base offers the best hope for reviving the Russian economy. To some extent, this is a false perception, promoted by decades of work in conditions of secrecy and isolation from the outside world.

Another belief, which has very deep roots in Russian culture, is faith in the "cosmic" destiny of Russia. The massive Soviet commitment to space research was probably reinforced by such beliefs and now, at a time when funding for space technology is under severe pressure, there remains a strong conviction that it is Russia's destiny to be at the forefront of space exploration.

Given the profound problems of the Russian economy and the urgent need to improve the population's living standards, health care, and environment, the country's leadership faces a serious policy issue: to what extent can Russia afford to retain a substantial commitment to frontier technologies in military, space, and civilian fields? Aspirations may no longer be compatible with material possibilities; a re-ordering of priorities should now be high on the policy agenda.

Within the Russian scientific community one also finds a deep-rooted conservatism and unwillingness to change, which has been apparent in recent years in leading circles of the Russian Academy of Sciences. This conservatism has been reinforced by the habit of state support and the expectation that, in the last resort, the State will always find the resources to keep research institutes and activities alive. It will probably take many years to overcome this "dependency culture". The traditional forms of state support also fostered a social psychology of levelling – the view that available resources should be distributed evenly among S&T organisations – and this makes the adoption of selective, competitive forms of funding involving peer review difficult.

Another unhelpful bias of the Soviet regime was a tendency to assume that the larger an organisation the better. Such "giantism" was reinforced by the endemic administrative logic and, in relation to research activity, largely unconstrained by consideration of diseconomies of scale. Russia has many research establishments that employ thousands of staff and often have serious problems of internal bureaucracy.

6. Concluding remarks

In the present difficult circumstances, it is understandable that issues of the short-term survival of the S&T base are a major policy concern. The Russian Government is faced with the problem of combining short-term measures with policies for creating an S&T system appropriate to a market economy and a democratic society. Tensions are inevitable, as action to halt negative trends may require resorting to administrative actions or other measures that run counter to the process of market transformation. In conditions of flux and uncertainty, it is also difficult to identify genuinely negative and positive developments, a difficulty exacerbated by powerful bureaucratic interests able to make vigorous claims for special short-term support. In principle, it is evidently desirable that, whenever possible, short-term actions should support the processes of longer-term transformation. The challenge now facing the Russian Government is to secure the appropriate mix of measures required.

While foreign assistance has an important role to play, in terms both of collaboration and technical aid, the problem of transforming Russian science and technology is, above all, a matter for the Russian Government and people. The limitations of external aid must be recognised. The best condition for positive, ongoing, involvement of foreign partners is domestic political and institutional stability.

Chapter II

The policy framework

In Russia's present difficult circumstances, the S&T policy framework and the control of S&T policy are naturally strongly affected. A number of factors make the problem of formulating and guiding S&T policy particularly complex:
- important changes have been made in the structure of command, at the level of the state in general and at the level of S&T policy co-ordination in particular;
- as noted above, the central government's power has eroded considerably, and economic and political power has gradually shifted towards regions and enterprises;
- real power now depends more on the funds at the disposal of institutions and on their access to hard currency than on formal responsibilities, as defined by laws and decrees.

Despite the uncertainties engendered by the present situation, the public authorities have taken a series of measures, most of which seem to go in an appropriate direction. This chapter addresses issues relating to the organisation and management of S&T policy as part of the more general changes in organisational structures. It examines the funding of S&T in the light of these changes and the current priorities for its funding. Finally, it discusses the role played in the evolving picture by international S&T co-operation.

1. Guiding institutions: from the old system to the new

Political co-ordination

Under the Soviet regime, science was essentially guided by the Communist Party network organisations, through their co-ordinating bodies at the central, regional and institute level. These bodies supervised academic, university, and industrial R&D, which were treated as distinct entities, and they assured whatever co-ordination was necessary among the various S&T areas. While the Academy of Sciences played a key role in developing basic research, industrial and applied R&D were largely oriented towards the military sector, and the main responsibility for industrial R&D lay in the hands of industrial ministries. These were under the command of the Military-Industrial Commission (VPK). Through such "political" networks, which operated both vertically and horizontally throughout the whole system, S&T was effectively guided and controlled, and information was regularly and accurately channelled to policy makers. With the suppression of the Communist Party, these networks have disappeared, at least formally. However, the tendency to maintain similar structures remains to a certain extent, notably in the regions.

Central bodies in charge of S&T

At the central level, the State Committee for Science and Technology (GKNT), chaired by a high-level government official with the rank of minister, was formerly responsible for national S&T programmes. It was also responsible for international co-operation, and in fact offered Soviet scientists their main opportunities for contacts abroad. It had links to intelligence organisations and was closely supervised by the VPK. This Committee was the basis of the newly established Ministry for Science and Technological Policy (Minnauka), which is now responsible for the planning of the entire civil R&D effort. Minnauka directly controls only part of the funding, but it has some influence on R&D expenditures made elsewhere. Like its predecessor, it is in charge both of the national S&T programmes and of international collaboration.

Advanced technical institutes and universities (both subsumed under the Russian term "higher school") were formerly under the responsibility of the State Committee for Higher Education. University research was very limited due to the strict separation of science and education. When Minnauka was formed, it took over the responsibilities and staff of the Committee, but higher education was later separated from Minnauka (March 1993), and a new ministry was created.

Major ministries

In the past, the Ministry of Defence was an influential policy maker for the S&T system, and it continues to play a strong role. A post of First Deputy Minister for Science has been established (it is occupied by a civilian). Among the sectoral ministries, the most influential were and remain the ministry in charge of nuclear weapons and atomic energy (Minatom), which controls all related scientific research, uranium mining and production, and nuclear power stations, along with the ministry in charge of the missile and space industry, part of which has become the Russian Space Agency (RKA). Other ministries, now disbanded, controlled industrial R&D and production. A State Committee for Industrial Policy now oversees the industrial sector, aided by committees for the defence, civil machine-building, metallurgical and chemical industries. In addition, the Ministry of Finance, in charge of the overall national budget, plays a background role in S&T funding, as it has a say in allocations to S&T, which have also been affected by the economic policy reforms put forward by the authorities.

The Academy of Sciences

The Academy of Sciences dates from the 18th century and is a unique body, conceived not simply as an honorary society of academicians but also as an active network of research institutes. Its more than 300 research institutes employ about 12 per cent of the R&D labour force (see Background Report for further details). During the Soviet era, in a context of great respect for science, the Academy grew. It was a strong, rather independent, and politically neutral organisation, which formed a kind of "scientific republic" within the Soviet Union and had significant influence on scientific policy formulation and policy setting. Today, with the general decline of the status of science, it has lost some of its past influence. With the breakup of the Soviet Union, the Academy lost important institutes in Belarus and Ukraine, while in Russia several institutes have tried to leave the Academy system in the hopes of receiving increased funding from other sources.

During most of the existence of the Soviet Union, the Academy of Sciences was largely, but not exclusively, a Russian institution. In 1990, serious discussion on establishing a separate Russian Academy began. However, after the planning had progressed quite far, there was a complex series of negotiations and reorganisations, as a result of which the embryonic Russian Academy and the existing USSR Academy were combined to form the current Russian Academy of Sciences. The basic outlines of the Academy's structure have been drawn, but many aspects of its organisation, operation, and legal status remain unresolved. Two other academies that existed previously and continue to exist are the Academy of Medical Sciences and the Academy of Agricultural Sciences.

General comments

Although these policy and organisational changes are not considerable, they call for the following remarks:
- With the disappearance of the formal top-down mechanisms for political co-ordination of the Communist Party system, there is now a certain vacuum. New mechanisms are needed at the highest government level. An interministerial commission has been established and is chaired by the Minister of Science and Technological Policy. It met for the first time in July 1993 and brought together some 60 representatives from the relevant ministries. It seems advisable to consider the possibility of co-ordination at the level of the Prime Minister, with the participation of all key partners, including defence and finance ministers. (In addition, it is worth noting that the President's office has a Security Council and a Science Advisor; this creates a double power structure and also raises questions of co-ordination, as conflicts with possible co-ordination at Prime Minister level may arise.) If all ministers involved (even peripherally) with S&T participated, it would be possible to institutionalise innovation policy and the management of the innovation system as these concepts are now understood in the OECD countries – a concept much broader than science and technology in the strict sense.
- Minnauka seems the natural tool for elaborating and implementing S&T policy as such, and it should be responsible for preparing the civil R&D budget and administering the federal R&D programmes. While this may present risks of bureaucratisation and centralising control, a decision-making body which can arbitrate between the various corporatisms that

are part of the S&T scene is needed. It should develop management methods in line with those of OECD countries, notably as regards the administration of funds. Minnauka should work to rebuild the international network of scientific attachés, which has largely disintegrated, and assure its co-ordination. It also needs to establish close relations with the new patent organisation and the state committee in charge of standards and certification.

- It appears reasonable and natural that the Academy of Sciences continue to manage its network of institutes and maintain leadership in the supervision of fundamental research throughout the country. The Academy-dominated research system, separate from the university system, was peculiar to the Soviet approach (and was also imposed on its satellites). The approach may not be the best (see Chapter III), but Russia's scientific strengths lie there today, and, to some extent, the Academy is irreplaceable. On the other hand, it is advisable that those responsible for the universities be more and more involved in the management of basic research. Moreover, the Academy, like other bodies, should be publicly accountable, as far as possible, for the use made of funds received from the State.

- In addition, it is worthwhile mentioning the creation of a new institution, the Russian Foundation for Fundamental Science, formed in 1992 by Presidential Decree, which is to support research projects in basic science in a manner modelled on the OECD country practice of using peer review. It was initially placed under the leadership of one of the vice-presidents of the Russian Academy of Sciences. Under pressure from the Minister of Science and Technological Policy, leadership has been transferred to another Academician, who is not a member of the Academy Presidium, to avoid possible conflicts of interest. While the Academy should certainly be closely involved in the management of the Foundation, it would seem appropriate, and in the Academy's interest, that the Foundation nonetheless remain an independent body, under the general administrative supervision of the Ministry of Science and Technological Policy (as is the case for similar agencies in OECD countries).

2. New actors

Parliament

The debates and the struggles that have taken place between the Government and the Parliament have slowed reforms in science and technology as in other important sectors. On the other hand, through its Commission for Science and Education, the Parliament voted important laws on intellectual property (autumn 1992), thereby showing that, through appropriate initiatives, Parliament can usefully contribute to S&T policy making by stimulating needed changes.

It is, generally speaking, good that a legislative branch, duly elected in a fully democratic manner, has an important voice in determining policy. However, when questions of tax policy, resource allocation, property rights, and other vital issues are subjected to legislative bargaining, difficulties, often in the form of conflicts of interest, inevitably arise. For example, overall endeavours to ensure or preserve regional balance can be countered by efforts on the part of powerful legislators to steer projects to their own constituencies.

Regional and local authorities

Weakening of the central government has resulted in a strong rise in local and regional initiatives, occasionally accompanied by tendencies towards autonomy. This effectively tends to mean local and regional setting of S&T priorities. Although this movement contributes usefully to removing a perennial dualism between the Russia of the capitals (St. Petersburg and Moscow), which has been far more open to the outside world, and that of the countryside, which has remained inward-looking, it is limited by the fact that 45 per cent of Russia's scientific potential is found within a radius of 100 kilometres around Moscow, while 15 per cent is in the St. Petersburg area.

Structures for developing regional S&T programmes proliferate, aided by a local self-organising process. In Moscow, the RENATEKS (Regional Scientific Technology Centre) has been established to focus local S&T efforts and collect funds from both central and local bodies. In St. Petersburg, a regional centre to support innovation has been established with the help of local and central authorities (Regional Foundation for Scientific and Technological Development); the leaders of large research institutes have met in order to select a few priority areas and involve foreign investors. In Nizhni-Novgorod, under the leadership of the vice-governor, teams from various bodies (RAS institutes, Minatom, universities, local branches of the Standards Committee, etc.) have joined together to serve local needs as well as the overall Russian market. A number of initiatives have also been taken by formerly closed cities specialised in specific technologies, in order to exploit their competencies in the embryonic market economy that is taking shape (*e.g.* Obninsk, Saratov, Krasnoiarsk). The experience of OECD countries in the field of innovation and technological development suggests that this movement should be encouraged.

Minnauka is helping this process with some funding and has selected a dozen sites, in principle to favour the development of centres modelled on Moscow's RENATEKS. The Ministry and the Academy of Sciences have established a joint commission for monitoring R&D regional development, and a similar joint approach with the Parliament is evolving. More generally, Moscow-based ministries and agencies should install representatives and offices in all regions so as to facilitate implementation of centrally determined policies, whatever the decentralisation trends may be.

The construction of science parks and technopolises is also often proposed – in the ministries, the Academy of Sciences, and the university – as a solution to financial and organisational difficulties. This "technoparktopia" recalls other large-scale concepts in Soviet S&T administration, such as the scientific-technical revolution, science-intensive industry, and science-industrial complexes. However, science parks will only be effective if solutions are found to problems of financing and inter-departmental co-operation.

Society and pressure groups

Russia has a weak tradition of free pressure groups either in politics or in science. Now that the S&T structure has been split up and the different parts are physically separated and have to fight for limited funds, creating real pressure groups is difficult.

One of the most influential non-parliamentary pressure groups is the Russian Union of Industrialists and Entrepreneurs, under the leadership of A. Volsky. Its members include the directors of many of the country's largest civil and military enterprises and also representatives of some foreign companies operating in Russia. It maintains close links with counterpart organisations in other New Independent States. In the military sector, another influential pressure group is the League for Support of Enterprises of the Defence Industry.

In addition, free "academies" have mushroomed. Here as elsewhere, there is inflation, and there are about 40 of these new academies, largely formed to give prestige to their self-appointed academicians. However, some are important and respected organisations (Technological Academy, Engineering Academy, Academy of Natural Sciences). They can play a role as horizontal linking organisations, and they are business-oriented, active in international co-operation, interested in fund-raising for projects, and, occasionally, conference organisers. They might also be the beginning of trade unions. One new union has been created, the Trade Union of Personnel of the Russian Academy of Sciences, but it is difficult to say how much influence it has at the political level.

3. The science budget

Given the current penury and shortages, the power of the different actors depends more clearly on the money they control than on organisational structures. This makes it all the more important to examine closely the science budget and the ways in which it is allocated.

The science budget as a whole has declined as a share of the total government budget. It represented 7.25 per cent in 1990 and only 3.4 per cent in 1992, according to estimates provided in the Background Report. Some 60 to 70 per cent of total R&D is estimated to be financed by the defence sector and 30 to 40 per cent by the civil sector. The defence sector is estimated to perform 30 per cent of civil R&D in its total research activity. Since 1989, the volume of military R&D undertaken by the defence complex has been reduced by more than one-third, with a reduction of about 18 per cent in 1992. Budget-financed civil R&D in the defence industry was cut by almost 40 per cent in 1992. Thus, the overall civil R&D budget seems to have diminished more than the military one.

In 1992, the civil R&D budget amounted to 102 billion roubles. At the beginning of 1993, the budget was to be some 150 billion roubles, but the sum was increased to some 700 billion in late March to keep up with hyperinflation.

This increase does not appear sufficient to match the pace of price increases; moreover, there are important losses of purchasing power because credits are only transferred on a quarterly basis. The sole remedy for this situation is a drastic reduction in the rate of inflation.

Data provided by Minnauka indicate that the allocation of the civil R&D budget can be sketched out as follows. The academies and branch ministries receive 60 per cent as institutional funding, and the RAS obtains 25 per cent of this sum as a lump sum, which the Praesidium allocates as it chooses to its institutes (in the current economic situation, most goes to salaries). Most of the remaining 40 per cent of the civil budget goes through Minnauka, which is responsible for the planning of civil R&D, to a number of actions related to programme funding. The most important of these items are:

- the national priority programmes (18 per cent of the total civil R&D budget) of which the space and aviation sectors receive 40 and 30 per cent, respectively, the corresponding sums being administered directly by the relevant ministries;
- the federal research centres programme (9 per cent of the total civil R&D budget), the purpose of which is to "save" some 30 top-level centres, mostly branch institutes with advanced competencies and a few located in universities; Academy-related institutes have refused to participate in the scheme, but their attitude seems to have evolved.

The new Russian Foundation for Fundamental Research receives 3 per cent of the total civil R&D budget and is, at present, the principal source of funding for new projects in basic science; some 80 per cent of the allocation goes to Academy research teams.

It is worthwhile mentioning, in addition, that during the period of tight economic policy in 1992, the Finance Ministry reportedly strongly opposed demands for budget increases from Minnauka and other S&T-related ministries. The Parliament has also affected the budget in general, and that for science and technology in particular; the latter has greatly suffered from untimely cuts due to sectoral demands (for sports, agriculture, etc.), although it is of course not inappropriate that the Parliament exerts some control over the use of funds.

As a whole, the distribution of civil R&D funding – which includes a relatively important share of flexible "free" money not tied to specific institutions, at least on paper – seems to offer fairly good opportunities for helping to restructure the R&D system. It remains to be seen how it is managed in the real world on a day-to-day basis. This is largely a question of the discipline exercised by the actors. It is clear that certain institutes receive privileged financing on the basis of former network contacts. More problematic is the rumour that a number of institutes (or even research teams) lobby the Ministry of Finance and succeed in obtaining funding directly. If that is the case, it will be impossible to develop a coherent science policy, and the disintegration of the whole system will become inevitable.

A Russian Technology Development Foundation has been established as a network of 150 branch funds and one federal fund, with resources derived from levies paid by industry (1.5 per cent on sale prices of products), with a view to supporting the federal share of costs of R&D equipment and the social sciences. However, only 10 per cent of the expected funds have been collected.

In addition to direct support mechanisms, the Government provides a considerable number of tax exemptions to scientific institutions (*e.g.* from VAT, from land property tax, etc.). Similar measures apply to new science-based firms which the Government is attempting to support. The real impact of these incentives and additional sources of money is unknown. For the time being, with only an incipient and essentially uncontrolled tax system, it seems ill-advised to make great use of tax mechanisms as policy tools.

The privatisation process is another source of funds for which a number of government circles have high expectations. As the rules of privatisation remain unclear, it is necessary to define what can be privatised and what should remain state property as basic assets for the national science effort. So far, unauthorised privatisation has taken place in a number of institutes. The Parliament has voted a law forbidding the sale of Academy institutes' assets, but privatisation continues, and institutes also often rent part of their premises to earn money.

Finally, the future of regional and local S&T programmes will depend heavily on the money they receive from local and regional governments. This will ultimately be conditioned by the revision of the taxation system, which currently does not formally allocate anything to local authorities, yet local enterprises and institutes are asked to support housing and social needs.

4. Priorities

Although important efforts are made to optimise budgetary allocations, priorities depend largely on established power relationships among institutions and on ideological trends. The *de facto* priorities can be sketched out as follows, including the defence sector, with its own priorities, as an important part of the whole picture. This assessment is not based on clear statistics provided by Russian authorities, but rather on trends commonly perceived by the different actors (mid-1993).

- R&D directly related to military activity probably still receives – in relative terms – substantial support. Recent high-level discussions in Russia stressed the need for more weapons exports as a source of hard currency. It is thus probable that related R&D continues to be a high priority. Nuclear research related to defence and energy production also continues to benefit from good support. Space activities developed by the space agency (RKA) receive some 6 per cent of the total S&T budget.

- Applied and industrial R&D for civil purposes is probably the area that has suffered most from current restrictions. Support provided by the military sector has been cut, and funding from branch ministries has largely dried up. Cuts in the aviation sector have been compensated, to some extent, by funding which is relatively generous compared to that for other areas in the national priority programmes.

- Basic research pursued under the auspices of the Academy of Sciences has been somewhat protected, despite relatively severe cuts. Support provided by the Russian Foundation for Fundamental Science has offered the opportunity to finance new activities. As a whole, the Academy has improved its relative position from 1992 to 1993, as it should receive about one-seventh of the civil R&D budget instead of the one-tenth it apparently received in 1992.

- The traditional disciplines such as physics, chemistry, and the "technical sciences" used to receive sizeable funding and continue to do so; the biological and life sciences were, and still are, the "poor relations" among the scientific disciplines. The social sciences have lost most of their institutional support and the field is being deserted by researchers.

A list of overall priorities has been established at the central level, and it resembles that of other countries throughout the world. It includes: industrial technologies, informatics and telecommunications, new materials, energy, transport, natural sciences and biotechnology, and ecology (see the Background Report). The actual priority-setting process at a more detailed level remains obscure. No less than 40 000 groups of experts are supposed to contribute to determining priorities and evaluating projects and institutes to be supported. The criteria used for selection are unclear. There is, unquestionably, an inclination to distribute small amounts of money to a large number of teams, and there is still a tendency to select frontier technology programmes without giving sufficient attention to their applicability.

A final remark concerns the neglect of funding for S&T services which are crucially important for the diffusion of technology or the exploitation of research results, such as patenting, standardisation, and quality control of products. The relevant organisations lack the funds necessary to fulfil their tasks.

5. International co-operation

S&T co-operation with Russia is generally and chiefly motivated by the very high prestige of Soviet science, particularly basic and theoretical science. Its remarkable achievements in a number of areas have always aroused the greatest interest in scientific communities throughout the world, and attempts to extend collaboration naturally followed the opening of the country to the outside world.

Another motivation is the desire to prevent the loss of Russian S&T – particularly the know-how, techniques, and research personnel in the areas of nuclear, biochemical, and military technology – to "unfriendly" countries of the developing world seeking to strengthen their armament. More generally, there is a desire to maintain S&T in Russia as a basis for economic development in the longer term, although some groups in OECD countries and some Russians feel that certain OECD and non-OECD countries have indulged explicitly or implicitly in the pillaging of Russian science and technology. Russian claims of "brain drain" are sometimes viewed as overstated, given that the bulk of it is to other sectors of the economy, and interpreted as a means of attracting greater interest and funding from abroad.

The New Independent States (NIS) are becoming a significant focus of collaboration and co-operation, with Russia accounting for some 80 to 90 per cent. The magnitude of present co-operation and collaboration with Russia is extremely difficult to assess, but informed sources in several countries estimate that programmes have increased some 30 per cent over the last two years (1991-92). As a whole, one may estimate that about one billion dollars are currently budgeted in OECD countries for S&T co-operation with Russia when both the public and private sector are taken into account. Of this, only a few hundreds of millions of dollars are directly spent in Russia, a sum which is not negligible, however, at the current rouble/dollar exchange rate (see Box B for further details).

Box B

What is the extent of scientific and technical co-operation between OECD countries and Russia?

Precise evaluation of the amounts invested by OECD countries in Russia is very difficult. Public budget allocations are often included in programmes for all the countries of eastern Europe and the NIS. In many countries, S&T co-operation is included in co-operation in education and culture, while co-operation for space and the nuclear sector is often part of military budgets. While many agencies and ministries are involved, there are generally no co-ordinating and centralising bodies. Many countries contribute to multilateral programmes such as the International Science and Technology Center (ISTC). In addition, initiatives of research institutions and universities are paid for directly, as are actions by private foundations and associations. Finally, the private industrial sector invests not inconsiderable sums, although the actual amounts are unknown. For all these reasons, it is only possible to make very gross estimates.

For public budget allocations, excluding sums budgeted for nuclear safety, the OECD Secretariat estimates the following amounts for the major G7 "investors" for 1993: United States, $200 million; Germany, $40 million; Japan, $30 million; France, $30 million. Budget allocations from all other OECD countries are estimated at $150 million. To that figure should be added funding provided for in multilateral initiatives – ISTC and the recently created EC Association for NIS scientists – which should represent approximately $30 million for 1993, as well as funds provided by the International Science Foundation (Soros), estimated at $20 million, and by international and national scientific associations (in physics, astronomy, biology, etc.) estimated at $10 million.

(continued on next page)

(continued)

This gives a global sum of approximately $500 million. In fact, part of these amounts: 1) will not all be spent because too few adequate partners or interesting projects are found in Russia or because of administrative problems in the donor countries or in Russia; 2) will be paid to firms or laboratories in donor countries and never reach Russia, even if the projects carried out are wholly directed towards Russia; 3) will be paid in the donor countries in order to welcome Russian researchers and technicians, who spend most of what they receive in the host country. In fact, the sums effectively paid to, and spent in, Russia are probably of the order of $100 to $150 million. Amounts spent by private industry are conservatively estimated to be of the same order. Thus, in 1993, $200 to $300 million should reach Russia and should directly benefit the Russian research system and the economy.

However, part of this amount enters the country secretly, part is paid by government agencies and firms for their travel expenses, and finally, even in the case of completely official and legal contracts, part goes directly to individual researchers or teams, without the knowledge of the ministerial and other authorities in Russia, and even sometimes those of the institutes. As a result, the authorities see far less than what is announced by the OECD countries.

Clearly, if the figures indicated above are exact, this is an important source of leverage for influencing the Russian R&D system, since, at an exchange rate of 1 000 roubles to the dollar, foreign funds effectively spent in Russia (in the form of hard currency) represent some 15 per cent of the total Russian R&D effort, including that of the military sector.

The OECD Member countries use a variety of mechanisms to develop co-operation and collaboration with Russia. The most common is the exchange of personnel and visits by Russian scientists. OECD countries are moving towards placing greater emphasis on targeting not only institutions and subject fields, but even specific researchers for specific projects. Another form of collaboration is the twinning of research institutes for exchange programmes and joint projects. Twinning agreements are particularly aimed at fundamental research and largely concern perceived centres of Russian scientific excellence. Joint projects are an increasingly important mechanism, whereby scientists or institutes in OECD countries identify projects for joint work with a specific institute or team in Russia and seek funding from national or multilateral programmes. Programmes of colloquia, conferences, summer schools and the like are seen as important means of building networks between Russian and foreign scientists, as are expert missions sent to Russia to identify areas of potential interest for collaboration. In addition, a number of research institutions have placed their own experts in Russia to help develop joint projects and to assist in information gathering and assessment. Their efforts complement those of science counsellors, who also fund some aspects of collaboration.

Private foundations have recently become more actively engaged in co-operation efforts in the Russian Federation, largely by funding projects or researchers. The most important of these, the new International Science Foundation created by George Soros, awards grants to the Russian scientific community [$100 million are to be spent for this purpose over a three-year period; after five months of operation (March-July 1993), it was reported to have received 32 000 requests from NIS countries (25 000 from Russia) and to have supported 18 000 in the form of $500 individual grants].

Industry in OECD Member countries is attracted to Russia and other NIS for several fundamental reasons: the presence of a well-educated body of workers; the favourable conditions offered to foreign investors, when compared to those of other less developed areas of the world such as Latin America or South-east Asia; and, finally, the long-term prospect of a vast consumer market. However, investments made by most OECD country firms are expected only to bear fruit in the very long term (ten years or more), particularly in the case of R&D.

R&D collaboration take various forms, ranging from simple consultancy contracts, to partnership contracts of variable length and importance, to the establishment of joint enterprises. A form of scientific subcontracting, using Russia's low-cost science labour force, is developing on a broad scale. It is not unusual, and is becoming more frequent, to see businesses in OECD countries recruiting and financing teams of tens or even hundreds of high-level Russian researchers in specific institutes for a period of several years. Choices are generally made after an in-depth evaluation of the different institutes and of the teams working in all disciplines. While OECD country firms generally appreciate the high scientific level of their Russian partners, they find the level of technological competence relatively weak, except in the case of advanced military installations.

Structural problems within Russia render international co-operation difficult, although the problem is somewhat more acute for industry than for fundamental science, where collaboration has been practised for a longer time. Among the major problems for all collaborative efforts, some have already been mentioned: the political instability, the organisational turnover which makes it difficult to find signatories to agreements who can make long-term funding agreements, and the marked decentralisation tendencies in some areas. Russia's banking system is still insufficiently developed, and this makes transferring funds to Russian researchers a very complex process. Present heavy taxation and customs duties result in the loss of as much as 60 per cent of funds transferred. There is also the lack of effective intellectual property laws and of a legal framework for co-operation with private institutions. A further basic problem is the poor infrastructure and quality of life in Russia and the difficulties that arise from the lack of a shared language – not simply in the linguistic but also in the cultural sense. It is true, as well, that various external regulations impede collaboration, both multilaterally (COCOM) and within OECD countries (*e.g.* the United States).

At the same time, the Russian authorities, for their part, are far from satisfied with current trends in international co-operation. They observe that large amounts of planned (announced) budgets of OECD countries are apparently not spent, whatever the reasons given (impossibility of finding appropriate partners in Russia, etc.). They do not appreciate the fact that considerable amounts of money transferred to and spent in Russia remain hidden in order to avoid heavy taxes or customs charges (even if these "grey" transfers benefit government officials as well). And finally, and mostly clearly, they dislike the tendency of foreign agencies, foundations and enterprises to by-pass the co-ordinating mechanisms that exist at the level of central bodies in order to deal directly with individual scientists and research teams. The major argument behind this complaint is that these practices contribute to the disorganisation of the science system, do not help the pursuit of a coherent overall policy, and encourage the "dollarisation" of Russian science. They would prefer, for instance, that foreign support be invested in the maintenance of infrastructures needed for scientific work. To some extent, the disorganisation argument may be sound, and foreign partners, notably government agencies, should adapt their methods. On the other hand, Russian central bodies should show proof that the traditional bureaucratic and corporatist behaviour that foreign funders find inappropriate is really changing.

An important illustration of the difficulty of international co-operation can be found in the problems involved in setting up multilateral initiatives such as the Moscow-based International Science and Technology Center, with a budget of $80 million, which is to support scientists and engineers involved in military and nuclear research over a period of three years. The operation, funded by the United States, Japan, and the European Community, has required considerable ingenuity in the search for mechanisms of payment, for protection of intellectual property, for facilitating further technological exploitation on a fair basis between partners, etc. The project was blocked by the Russian Parliament (until it was dissolved in autumn 1993). Its approval was required, but it feared that the project, as it stood, would simplify the "stealing" of unique Russian knowledge and technology assets.

A final point, and one on which information appears to be unavailable or relatively uncertain, is the question of S&T co-operation between Russia and the other republics of the former Soviet Union. If, as will be discussed below (Chapter III, Section 1), the greatest share of the R&D infrastructure was located in Russia, Russia's overall R&D framework has nonetheless suffered from the breaking up of the system. Ties among members of scientific and industrial communities have been maintained, but they do not suffice, on their own, to reconstruct the necessary links. In addition, some republics – notably Ukraine, which has several thousand nuclear warheads and a non-negligible part of the former military-industrial complex – have maintained a strong position in military technology.

6. Concluding remarks

Over the long term, it can be expected that the gradual establishment of a democratic framework and the political stabilisation of the country will result in a good balance of the different viewpoints that need to be taken into account in the policy-making process. It is clear that Russia has embarked on the difficult process of building a "civil society" with appropriate political, administrative, and financial mechanisms for decision making and resource allocations. This process is the source of most of the problems evoked above concerning S&T policy setting. The measures taken so far seem to go in the right direction. Major issues and further progress may depend more on the establishment of broader framework conditions related to taxation, privatisation, and finance and on the sharing of responsibilities between regional and central authorities, Parliament, and the Government. New types of relationships also need to be established with the international community to facilitate S&T co-operation.

Finally, as regards the financial aspects, it should be stressed that the State now finances the quasi-totality of the national R&D effort. Industry's contribution is currently estimated to be less than 5 per cent of the total R&D effort. The State alone cannot ensure all R&D financing. Such a situation is not sustainable in the long term; it is also unwise from the viewpoint of efficiency. The development of a market-driven R&D effort by industry, however, will take a long time. General improvement in the economic situation is needed, new forms of accounting and cost/pricing methods have to be diffused, and a new innovation system with improved conditions for entrepreneurs has to take shape (see Chapter IV). In the long run, when the taxation system has been fixed and possibilities of tax evasion are reduced, it will be worthwhile establishing appropriate incentive mechanisms for investment in R&D and innovation. This will help encourage industry to support and rebuild an S&T structure, very large parts of which are likely to disappear in the coming years.

Chapter III
The science base

1. A profound crisis

The Russian scientific community is probably going through the most troubled time it has ever known. The first problem arose with the disintegration of the USSR. The Soviet Union's research system was both centralised and compartmentalised. Russia had a large share, but there was an integrated network which included the other republics. As a result, if about 70 per cent of facilities and specialists are now in Russia, this does not mean that R&D structures are intact. In some cases, the research facility is in Russia and its testing facilities or pilot production plants are in Ukraine or Belarus. It is difficult to know precisely to what extent the breaking up of the Soviet Union has damaged R&D capabilities. In certain areas closely tied to military needs and very largely located in Russia, such as aviation and space technology, the damage is probably limited. In others, such as biology or machinery, for which Russia had no more than 50 to 60 per cent of the research potential, it is considerably greater.

The second problem has been the financial crisis. Due to the very nature of the support it received under the Communist regime, the privileged R&D sector had no need to consider maintenance and development costs. As the science budget has been reduced (see Chapter II), the situation has changed markedly, and the entire sector has felt, to different degrees, the effects. The branch institutes, which represented some 70 per cent of the R&D sector in the late 1980s, have lost support from their former central sectoral ministries. Institutions in much of the research system must now pay for heating, lighting, repairs, and other infrastructure costs, which they previously received free of charge, at rapidly inflating rates and with the prospect of major cost increases in the coming years. Scientific institutions now are running on quarterly or even monthly budgets, a situation which makes conducting research very difficult. Often, funds, including salaries, are not distributed. The consequences of inflation and budgetary uncertainty are exacerbated by the virtual collapse of other sources of funding, with the exception of foreign support for recognised areas of excellence.

Up to the end of 1992, the personnel situation apparently remained relatively stable. In the face of unprecedented declines in GNP and a considerable loss of purchasing power as salaries failed to keep pace with inflation, the S&T labour force declined by less than 10 per cent per year over the period from 1990 to 1992. The pace seems to have accelerated in 1993. Those who have left have generally done so voluntarily, as Russians are, by tradition, distinctly reluctant to dismiss employees. At the same time, most of those who have maintained their institutional positions have taken on one or two other jobs to earn extra money. For reasons noted above, much of the drop in the work force has occurred in the industrial research sector and in the social sciences and humanities.

In the present circumstances, it is not only necessary to engage in an extensive process of scaling down, it is also necessary to solve complex restructuring problems. With respect to the performance of R&D, the large number of R&D institutes in the Academy of Sciences' network and the underdevelopment of university-based research create a serious imbalance in the system, which affects the training of researchers and, more generally, the dynamism of the higher education sector. Research and education will have to be better integrated. Approximately 6 per cent of the science labour force is found in the universities, as compared with twice that amount in the Academies. The situation in the universities and the technical institutes (higher schools) presents a double image. Their activity currently appears relatively stable, due to the fact that two-thirds of their budget comes from state funds. On the other hand, more than 30 per cent of teaching (and scientific) personnel are reported to have left the sector over the last three years.

The crisis has provided a unique opportunity to make major changes, but with so much apparently in jeopardy, it has also generated a desire to preserve as much as possible. No one wishes to take unpleasant decisions to cut back staff or reduce the size of institutions, and, indeed, it is not clear who can or should determine how many institutes should be "kept". Russian political and scientific leaders have demonstrated a fully understandable reluctance to confront and then enforce these unpalatable and politically difficult choices. As a result, however, the process is taking place haphazardly in response to market forces, without sufficient planning or attempts at rational resource allocation.

Whatever happens, the issue of training young scientists will present a special problem. Without reform, the entire science community will suffer from inadequate financing and will stagnate, and science will become less and less attractive as a career. If the economic shock is allowed to do its work unfettered, the number of institutes will decrease and the number of positions at those that survive will decline sharply. This means very few attractive places for young researchers, who, as the most mobile segment of the population, are the most likely to move to other pursuits or to go abroad.

2. The Academy of Sciences

A new Academy Charter is currently being discussed. In the existing climate of uncertain legal and economic conditions, some of the key issues remain unsolved. It appears that the Academy will continue to exist as both an honorific and a research institution. It is now autonomous and owns its property. However, limits will be placed on its right to dispose of or commercialise scientific equipment and research premises. The majority of the funding for Academy institutions will continue to come from the State, making the Academy subject to the Government's overall economic capabilities, to internal budget battles, and to the policies of the Ministry of Finance. From 1990 to 1992, the RAS budget decreased by 2.5 times in real terms (according to the Background Report).

The Academy has traditionally been the centre of the best basic research carried out in Russia. Throughout its history, however, efforts have occasionally been made to incite the Academy to conduct greater amounts of applied research. One of the paradoxes of the current situation is that the Academy's autonomy frees it from pressure to perform applied work, but the level of financing necessary to maintain the full complement of its institutions and staff is no longer guaranteed. Applied and commercial activities offer among the few alternative sources of financing, and many institutes have turned to them in various forms in order to support their staff.

The Academy leadership (the Praesidium) is composed of senior administrators who have been extremely powerful, because they have controlled funding and the right to travel abroad. The creation of independent institutes and alternative sources of financing may lead to a new distribution of power. In the end, the degree of power that remains in the hands of Academy officials will largely depend on their ability to maintain their role as the major source of financing for basic research.

Among the alternative sources of financing that have recently appeared, one of the most interesting is the above-mentioned Russian Foundation for Fundamental Research, which was established to award research funds on the basis of competitive peer review and is administered by members of the Academy. It held its first competition in early 1993 and has announced some 3 000 awards. The administrators have been criticised for the rapidity of the review process and have been charged with favouritism. The Foundation's administrators deny this charge and pride themselves on having put a new programme in place very quickly, although they expect that in the future the process will become smoother. If its share of state funding for research increases, it might indeed come to play a prominent role. At the current level, the amount of support involved is inadequate to significantly improve the situation for most experimental science programmes. In addition, the manner in which the Foundation's administrators are chosen, and its position with respect to other institutions, should be clarified. The institution is, however, a promising example of rational decision-making for funding.

The ageing of the Academy work force also gives rise to concern. Rapid expansion of the Academy research system in the 1950s and 1960s, followed by more modest growth thereafter, has created a situation in which a majority of the personnel at Academy institutes are over 50 years old. Under present circumstances, the influx of young researchers has been severely limited, and many of the most talented and productive scientists are working abroad either temporarily or permanently. As a result, many institutes have only older researchers on their staffs. Efforts should be made to recruit young specialists, and changes should be considered in the present pension and retirement laws, which discourage senior personnel from leaving their positions.

The Academy will, inevitably, have to adapt and become smaller. These reductions will also mean further curtailment of the administrative apparatus. In order to avoid destructive unplanned changes, the Academy leadership should develop a rational programme to scale down numbers of institutes and personnel and to make proposals for better integration with other scientific and educational institutions.

3. University research

One of the most widely discussed options for renewing the system of basic research involves better integration of the Academy and institutions of higher education. In the past, very few Academy researchers taught at the undergraduate level. A larger number, but by no means all, were involved in supervising graduate students. Given that the financial situation of universities is less precarious than that of research institutes, it would seem reasonable to have talented researchers teach in universities. While there are examples of such arrangements, the fact that the same instances are repeatedly cited (Pushchino, Novosibirsk, St. Petersburg Technical University) suggests that the initiative is not widespread. One problem is the perceived lower status of teaching; another is the lack of teaching experience among research scientists.

Practical measures to facilitate this reorientation might include:
- Programmes of the Russian Ministry of Science for providing additional research support to researchers who also teach and to teachers who also do research. More generally, universities and technical institutes should receive gradually increasing support for research work, especially since more than half of Russian doctors and candidates of science are employed in the higher education sector.
- Training seminars to assist research scientists to develop skills as university pedagogues, encouragement of "sister university" programmes that provide assistance to research scientists willing to undertake teaching careers, and exchanges that target university-affiliated researchers for sabbatical visits to OECD countries.
- Creation of a group of prestigious teaching institutes, for which the teacher/researchers are recruited at the highest level from the university system, the Academy of Sciences, and foreign countries (such as, for example, the Collège de France). The creation of a body of state professorships is currently being discussed. These professors, several thousand strong, would receive substantial salaries, paid by the Ministry. Such

measures might help bridge the gap between the Academy and the university system and strengthen the links between research and teaching.

The State Committee for Higher Education has outlined its own programme of scientific research, "Russian Higher School". Despite the relatively stable situation of the universities and institutes, this programme appears overly optimistic, as it promises financing to virtually every field of science and every institution. As for the Academy, the difficult choices and inevitable reductions will have to be faced eventually. They will be particularly necessary for the Higher School laboratories that previously carried out research on a contract basis with financial support from enterprises. Now that enterprises are no longer required to spend a specified portion of their funds on contract research and are, in addition, in serious financial difficulty, little money is being allocated to contract research. It seems unlikely that these research laboratories can become self-financing in the near future, and, unless educational institutions act quickly, the costs of the unutilised research capacity will pose a serious threat to their budgetary situation. With shrinking state financial support, the administrators of educational institutions, like those of research institutes, will have to reorient their thinking and seek alternative means of funding. A possible source of longer-term support might lie in the direction of serving local and regional needs.

4. The branch institutes

In the transition to a market economy, the fate of the extensive network of branch research organisations presents an especially acute problem. In the past, each branch ministry for a particular industry, or for non-industrial economic sectors such as agriculture, transport, and construction, developed its own research institutes, project organisations, and design bureaus; these owed their existence as much or more to the administrative logic of bureaucratic interests as to the economy's real R&D requirements. Once established, institutes followed the characteristic extensive growth pattern, with the result that branch S&T organisations proliferated. They were often overstaffed and gave rise to substantial duplication.

The activities of these branch institutes were often cut off from production, and their research efforts were oriented, not by customer demand, but by plans determined at higher levels of the administrative hierarchy. They were rewarded less for implementing research findings than for completing reports on work undertaken. Core institutional funding was guaranteed by the State, but from the late 1960s ministerial S&T funds created from levies on enterprise profits became increasingly important as a source of funding for civil branch R&D.

Branch R&D is suffering both from drastic reductions in state budget funding and, following the abolition of the branch ministries, from the loss of ministerial funds as well. As enterprises are now concentrating on short-term survival, they have no demand for R&D, nor, with sharply reduced profitability, the funds to finance it. Much equipment lies idle. Institutes and design organisations retain their staff by resorting to commercial bank credit or, when possible, by renting their premises. Salaries have plummeted, and large numbers of staff are leaving, especially the younger enterprising personnel able to exploit new opportunities in the non-state sector.

The branch S&T system will necessarily contract, possibly very substantially. At present, institutes, research teams within institutes, or individuals able to adapt to the new circumstances are devising ways to exploit opportunities that arise in the confusion of transition. If this process continues, it is likely to leave those unable or unwilling to change with little alternative but closure or unemployment. This spontaneous reorientation is by no means entirely negative, but it carries the risk that organisations capable of succeeding in more stable market conditions may disappear, leading to an unnecessary erosion of the nation's research capability.

Privatisation introduces an additional complicating factor. Very few branch research organisations can hope to survive as independent private establishments. In some cases, experimental production plants are being privatised, leaving institutes without test facilities. Science/production associations, formed in the 1980s, might offer a framework for the survival of R&D organisations if they can be privatised as united corporate entities. However, many of these associations may fragment as each member seeks to secure its own survival. Moreover, it has been proposed that a limited number of large industrial/financial corporations or holding companies should be created in key high technology sectors; they might provide an effective framework for strategically important branch S&T organisations.

If potentially viable elements of the branch S&T system are to be saved, the Ministry of Science and Technological Policy, the State Committee for Industrial Policy, and the State Committee for the Management of State Property will have to co-operate in order to identify strategically important S&T organisations, or parts of such organisations, for at least transitional state support. It is important that such intervention be very selective, and this will require procedures capable of overriding the residual power of the administrative logic of the old system. If at all possible, the instruments of intervention should also be compatible with market transformation.

As mentioned above (Chapter II, Section 3), the Ministry of Science and Technological Policy has elaborated a plan to "save" about 30 institutes considered as crucial national scientific assets. They are to have the status of "federal R&D centres" and receive priority support. The selection process appears to have taken place already (summer 1993). It is planned that this policy approach – consistent with foreign practice – will continue to be used.

5. Evaluating researchers, projects, institutes

There are three possible approaches to the evaluation of performers of R&D. In one, used for decades in the USSR, institutional administrators determine performance criteria and make judgements accordingly. While this procedure was plausible in periods of growth and even in times of stability, it is unsuited to conditions in which reductions in staff or elimination of institutes become unavoidable. This administrative approach was used, to some extent, to select the 30 institutes designated as federal R&D centres. The criteria applied were not performance-based; rather, they emphasised the excellence and uniqueness of the scientific "assets", with a view to saving Russian R&D. However, this approach presents serious risks of favouritism.

A second approach, used with some success in the former East Germany, Estonia, and Latvia, is to conduct a review of existing institutions in conjunction with outside experts, an approach which has the merit of apparent impartiality. However, the Baltic states are compact, with a limited number of institutes and universities. East Germany shares a language with the nation that absorbed it. In Russia, which is far larger and far more diverse, this approach may not be workable. To carry out a thorough evaluation of Russian institutes (similar to the process of attestation) would be an immense undertaking. However, this approach could be used for carefully defined sectors or sites.

A third method of evaluation is to establish competitive peer review mechanisms in order to distribute the available resources. In OECD countries, this method is mainly used to allocate grants for specific research projects. It is little used for institutional funding. Under current conditions in Russia, it may be applicable in both cases. There are a great many objections to reliance on peer review. Scholars in the Russian Federation have no experience with the process: they do not know how to write proposals, and many simply will not write them. There are no established procedures. The possibility that normative criteria will intrude is greater than it is in countries where the process is well-established. And the process itself is pernicious: it rewards the best proposal writers rather than the best scientists; it deflects time from real research to the search for grants; it gives no guarantee that time invested in significant research will be rewarded with additional opportunities to complete that work; and it puts too great a premium on immediate results.

All of these weaknesses are real, and all are well known everywhere peer review is practised. As yet, however, no one has devised a method for allocating public funds for research that better balances the competing interests of taxpayers, scientists, and policy makers. In some ideal world, everyone might work at his or her own pace with no demands for results. In the real world, and notably under current economic conditions in Russia, such luxury is clearly impossible. The sole question is what mechanism will be employed to implement the reductions, and the competitive peer review system seems the least undesirable option. Russian scientists already have, in fact, fairly good assessments of colleagues working in related disciplines and networks, and the recent experience of the Russian Foundation for Fundamental Research has shown that a peer review process can be implemented usefully.

6. Integrating Russia into the international scientific community

International co-operation is an indirect source of outside assessment, and it can help the Russian authorities to identify strengths and to select fields and institutes for priority support. Apart from well recognised ''poles of excellence'' (*e.g.* in mathematics), most OECD countries identify areas in the formerly closed military research sectors. These areas include: new materials, surface technologies, solid state physics, optical research, and propulsion technology. Opportunities for co-operation should be fully exploited, while appropriate protection and reward for genuine Russian knowledge and technology should be ensured.

In addition to obstacles of a general nature that affect collaboration between Russian scientists and foreign partners, noted above, certain specific difficulties related to research collaboration deserve to be discussed.

A major area of difficulty is the great variability of research capabilities and quality among research institutes and the lack of available information on the state and structure of Russian science and technology, especially when compared to the efforts of East European countries to make such information available. Some issues presented by Russians as scientific and technological – notably in agriculture and some areas of energy and production technologies – appear instead to foreign observers to concern infrastructure and sheer economic backwardness.

Another major problem is the absence of management skills in research establishments, in particular with respect to financing, external funding structures, and the mechanisms, such as associate laboratories and subcontracting, for ensuring collaboration of mutual interest with complementary objectives. A related issue is the lack, in Russian proposals for collaboration, of background information on the resources and the quality of science in the sector concerned, the internal resources available for the project, and the nature of the problems to be addressed. In addition, there appears to be a tendency to see all information, including what is elsewhere in the public domain, as being of value and for sale. Finally, the Russian science establishment seems to understand collaboration and co-operation in terms of aid exercises designed to maintain the existing science structure, whereas researchers in OECD countries, particularly in the private sector, seek projects of mutual interest. It is important that differences in understanding between agencies in OECD countries and Russian authorities be resolved.

There are areas, however, in which aid is perceived as essential on both sides: requests for support for infrastructure should, in particular, receive the utmost attention. Scientific work in most experimental disciplines has all but ceased due to lack of supplies and equipment. The dearth of hard currency at most institutions has made it impossible to obtain crucial materials from abroad or to maintain subscriptions to scientific journals. Scientists are increasingly unable either to conduct their own work or to keep up with the work of their colleagues. In a world where scientists exchange preprints of important articles, falling behind by a year or two is tantamount to professional obsolescence. The Russian Government appropriated $12 million in hard currency for the Russian Academy of Sciences to purchase foreign scientific literature in 1992, but no funds were actually provided. Instead, the Academy inherited a hard currency deficit of $175 million from the Soviet Academy. International and national associations of scientists (*e.g.* in physics, astronomy, mathematics, biology) have taken steps to provide needed scientific literature.

Other areas that deserve attention include:

- Often costly key materials and substances for conducting experiments should be provided. Technical research laboratories in Russia have always suffered from the lack of small but essential components and devices necessary for building prototypes, and this leads to considerable delays in conducting experiments. Ways to provide this kind of equipment should be examined. One possible solution might be to establish ''shops'' stocked by the international community.
- Donation of equipment. While all scientists and technicians prefer to have the latest technology, the Russian S&T community uses a range of instrumentation extending from the very latest to the near antique. A programme encouraging scientists in OECD countries to donate used equipment and supplies to Russian scientific and particularly educational institutions would make a major contribution.
- Support for communications, particularly telecommunications. This support should be part of a broader programme to develop and modernise the telecommunications infrastructure (see Chapter V), but the academic community would benefit from further strengthening and extension of existing networks (such as RELEARN). Russian universities are just beginning to develop electronic mail capabilities and greatly benefit from assistance provided in the form of equipment and training.
- Action is needed to save major facilities such as museums and libraries with unique collections. Some are deteriorating rapidly and immediate action is needed (as the US academies recently stressed in a meeting held

in Washington in February 1993 on "Sustaining Excellence in the Former Soviet Union").

Russia has a number of major big science facilities that could be used to benefit world science, *e.g.* in astronomy, oceanographic research or ecology (Lake Baikal). Specialised international associations could facilitate collaboration, as could the OECD Megascience Forum (in which Russia participates as observer and to which it has made proposals, as yet unanswered). One area in which collaboration seems to proceed relatively smoothly and at a significant scale is the ITER project on nuclear fusion; Russia is involved in the financing and management of a series of major test operations now being conducted throughout the world. Present world interest in research on fusion, the strength of the concerned scientific community at the international level, and Russia's recognised competence in the field explain why this has been possible.

7. Brain drain and brain waste

The financial crisis, the unpredictable food supply, and the general uncertainty make it very difficult for most specialists to continue their work, and the combination of economic stringency and professional frustration leads to substantial brain drain. There are multiple processes at work here. While most attention has been focused on scientists who have moved to jobs in other countries, emigration abroad has thus far been the least serious disruption. Laws from the Soviet era that prohibited Russian specialists who held security clearances from going abroad without special permission are still in force. The precise number of specialists who have emigrated is unknown, both because Russia is not fully aware of their movements and because some of the host countries refrain from making such information available (see Box C). The impact, however, is not measured by numbers alone, as the emigrants leaving are in many cases people with international reputations and with marketable skills. For example, in one Academy of Sciences biology laboratory this year seven of the nine laboratory directors were working abroad. More generally, it is to be feared that certain highly reputed scientific "schools", such as some in theoretical physics, will disappear, due to the dispersal of their personnel abroad.

Box C

How extensive is the emigration of Russian scientists and technicians?

This is a very poorly documented topic, for reasons that have to do both with the nature of the phenomenon and with the present situation in Russia. Two estimates are given in the Background Report, both of them based on data from the Russian emigration service. The first, provided by the department of the Ministry of Science in charge of social issues, takes as its point of departure the information that since 1990 a yearly average of 5 000 persons with advanced scientific or technical training officially declare their intention to emigrate and establish residence abroad. These statistics also include teachers and construction and transport engineers. The department estimates that this number must be multiplied by five in order to gain a more precise view of actual emigration and to take account of scientists who have gone abroad for periods of six months to a few years and who are temporarily employed in a foreign country. This would mean approximately 60 000 scientists and engineers over the past three years (1990-92). A second estimate, furnished by the Centre for Science Research Statistics, takes as its point of departure a more restrictive definition of R&D personnel and finds 2 000 to 3 000 emigrants per year. No estimate of a multiplier is given. The truth is likely to lie between the two, and total emigration (definitive and temporary) in a range of 20 000 to 30 000 persons seems a reasonable figure; it corresponds to some 10 per cent of the total loss of scientists and engineers from the research system.

Thus, several tens of thousands of persons are involved. A non-negligible proportion will in fact never return to their country of origin. Yet Russians are reputed to have a strong attachment to their home country, and comparative surveys with other countries of eastern Europe have shown that that less than 10 per cent of temporary expatriates seriously envisage remaining permanently abroad.

As for the countries in which Russian researchers have relocated, there are at present no reliable statistics. For the OECD countries, it is known that a large number have gone to the United States and Germany. Informed sources also mention significant departures to Korea, Mexico, and especially Israel, where, according to official statistics, 30 000 persons with advanced training arrive during every quarter; they are scientists, but also lawyers, architects, and doctors.

The Russian Academy of Sciences estimated that of its 1 701 researchers working (in 1992) on long-term missions (over six months) or under contracts abroad, 38.2 per cent are in the United States, 16.2 per cent in Germany, 8.9 per cent in France, 5.7 per cent in the United Kingdom, 5.2 per cent in Canada and 4.1 per cent in Japan.

The brain drain abroad does not have only negative aspects, however. Scientists who have emigrated can help integrate Russia into the international community by facilitating entry into expert networks, diffusing advanced knowledge and transferring new methods back to Russia. These emigrants constitute a valuable resource that should be exploited. In OECD countries, government agencies and universities should pay particular attention to establishing contracts and programmes that help high-level emigrants return home. Experience gained from practices such as twinning of institutions should be assessed from this perspective. The presumed inclination of Russian scientists to return home should also be assessed.

The more extensive internal brain drain involves the movement of individuals out of their fields of specialisation into activities that promise greater short-term material rewards. Salaries in the science sector were 30 per cent lower than the national average in early 1993, after having been 20 per cent higher in the late 1980s; these differences are, however, much less significant than the difference with the salaries offered in certain parts of the private sector, such as export-import or banking. Although changing occupation often entails geographic relocation and changing employers is very difficult because many aspects of personal life have long been tied to the workplace, growing numbers are doing so.

As talented individuals have increasingly sought employment in sectors outside their specialties, a number of enterprises have benefited, although there has been an influx into software development co-operatives and consulting firms, a sign of the general preference for employment in the service sector of an economy that still produces far too little. Nonetheless, it should be emphasised that the vast S&T complex and, in particular, the branch R&D sector is the repository of a large body of highly educated personnel who, beyond their traditional R&D activities, could contribute productively to a market economy. What in Russia is so often seen as a problem should be regarded as an opportunity, and, where appropriate, positive measures should be taken to facilitate redeployment.

Russia's unique complex of scientific institutions and personnel constitutes a human resource of world significance. Parts of the Soviet/Russian scientific edifice closely resemble those of other countries. Others are more idiosyncratic and may be as much a source of strength as of weakness. Still others cannot survive under current conditions. Different means can be used to address different aspects of the system. Thus, outside assistance can help those researchers and research groups doing work that can be subjected to peer review. Areas and state S&T programmes identified as priorities by the Russian Government are likely to continue to be funded, perhaps at a reduced level; these include some projects deemed important in Russia but of less interest to the international community.

Care must be taken to ensure that selection mechanisms are both varied and transparent and that instances in which individuals or groups obtain funding from multiple sources are minimised. This is a problem everywhere, but in OECD countries the governing principle is that an individual's full-time equivalent work time (FTE) must be honestly accounted for, and scientists who receive funding from multiple sources must inform funders of their situation. Safeguards will have to be adopted while a culture of accountability is being developed.

International assistance will require co-operation between Russian science officials and the governments and foundations providing assistance. Mutual trust, based on a close working relationship, is a necessary condition of success. Donors must be assured that their assistance will not be used to replace or reallocate funds from the state budget, and they must be certain that the Russian Government seriously intends to make difficult decisions and implement even unpopular reforms.

8. Support to young graduates

Science was once regarded as a highly desirable and prestigious career, but it is now considered an activity with little promise. Today, many talented young people are turning to other professions. In a country with an outstanding tradition of secondary school mathematics and science education, this represents a particularly serious loss.

Student complaints about the cost of living and inadequate stipends have been voiced from the beginning of 1992 and have increased despite rises in stipends, which inevitably lag behind inflation. Further, Higher School administrators do not know from month to month whether they will be able to pay the stipends. The Government is presently caught between the demands of unprotected groups in the population (students, but also pensioners and the unemployed) and the obvious need to control inflation. Thus, before the April 1993 referendum, the Russian President promised another round of stipend increases, but the Government later announced a delay in implementing the measure due to the need to control inflation.

It is also proving difficult to find positions for graduates, a situation that will inevitably deter talented individuals from continuing their education. The system of job assignments has broken down almost completely. The commissions that previously assigned graduates to three years of obligatory work now function as advisory bodies and clearing houses, but most employers are not in a position to hire new workers. Students are finding that employment in the private sector, and especially in commercial ventures, pays far better wages than work in which their education plays a direct role. It is even generally more lucrative to give private lessons than to teach in a state educational institution.

It is important to provide young S&T personnel with opportunities to work in other countries, but this must be done in a way that encourages their return to Russia. On the one hand, knowledge of English and integration into the international professional community enhance their effectiveness and make it easier for them to compete for research support. On the other, they also make it easier for them to emigrate. Under present conditions, there is little alternative to taking this risk.

The education system is an area of high priority for assistance from abroad. In addition to the actions concerning communications and equipment mentioned above, specific co-operative measures that should be adopted on a priority basis include:

– Programmes for graduate students and post-doctoral researchers to spend six months to a year in universities and laboratories in OECD nations. The danger that this will encourage brain drain is more than outweighed by the positive effects of exposure to a different scientific community. Many young Russian scientists have indicated that they would be content to remain in Russia if they have some assurance of regular travel abroad.
– Summer seminars, particularly for young researchers. The results of a small number of already developed programmes suggest that they provide an important source of psychological support, as well as benefits in terms of collaborative scientific work. It is a way to ensure that the next generation of scientists in Russia and in OECD nations are genuine colleagues. A number of such seminars could be organised in Russia itself.

9. The social sciences and the humanities

The social sciences and the humanities were the disciplines most severely affected by political interference, particularly in the Stalinist era. In the 1920s and 1930s, the social sciences were the main target of attacks within the Academy of Sciences. As a result, the generations educated during and after World War II were essentially deprived of these disciplines as they are understood abroad, although certain of them, such as linguistics, medieval history, and comparative culture studies, continued to maintain a high level and enjoyed an international reputation.

The social sciences and the humanities have also been the areas most severely hurt by the recent economic crisis. While almost all teaching and research personnel have experienced serious difficulties, the demand for social scientists has been even lower than that for individuals in the natural sciences and the humanities. Whether in the new commercial structures created in Russia and in some of the other new states, or in terms of opportunities abroad, social scientists are not in demand.

A massive programme of (re)training is clearly needed. Individuals who taught or wrote about Marxist economics, Communist philosophy, or Party history are not necessarily those best equipped to teach the skills needed in a market-oriented economic system. Yet these skills are in urgent demand. Basic business, accounting, marketing, and advertising skills are among the areas requiring greatest attention at present. Although some of the social science professionals from the earlier period may be able to adapt, many will have to find other areas of work. It might be argued that the need for state support is even greater for the social than for the natural sciences, which are likely to find it easier to obtain assistance from industrial, commercial, and foreign interests. It is worth emphasising as well that the social sciences and the humanities have an important role to play in monitoring economic and social change in the transition period. Moreover, many centres of creativeness – in the allied areas of the arts, cinema and literature – are also taking shape.

Finally, it should be remembered that product design is an essential aspect of innovation and an activity which requires familiarity with technology, understanding of the market economy, and the development of artistic skills. Particular effort should be devoted to developing this activity by establishing national or regional schools, with the support of the international community.

10. Concluding comments

In science and education, as in the economy, Russia is now at a crossroads. Rebuilding the old centralised state system is certainly out of the question, but the desire to preserve familiar structures remains strong and will continue to slow the transition to a new system. The alternative – moving rapidly to a highly uncertain, more diverse, and more dynamic new system – involves a painful and massive reduction. The only question seems to be whether the pain will be concentrated in a few years, or spread out over decades.

However, in proceeding with the necessary reduction and reorientation, it seems best to adopt a pragmatic approach that builds on the strengths of Russian science, identified through procedures such as peer reviews. If properly conducted, the process will lead to a research sector resembling the German one, in which the Academy plays to some extent the role of the Max Planck Foundation, the Japanese one, with important public industrial research laboratories, and the French and Anglo-Saxon ones with an important military sector.

Nevertheless, there are two important differences. First, the university research sector remains weak, and its connection with the rest of the system is inadequate. It will be important to follow closely its gradual integration with the Academy and to consider possible further reforms. Second – and this is a greater source of concern – the level of research within firms themselves has always been low, due to the separation of design bureaus, research organisations, and production facilities, and has dropped further. This is worrisome, especially for the nascent civil sector. To fulfil the needs of technology-based growth, powerful and efficient industrial research needs to be integrated in the firms themselves. For such research to develop, however, a market-based and demand-led innovation system must first be created, an issue that is addressed in the next chapter.

Chapter IV

The innovation climate

1. An embryonic market-based innovation system

For reasons that largely have to do with Soviet organisation but which also reflect Russian tradition, the introduction of an innovation system based on multiple, complex and continuous interactions between science and the marketplace will be slow and difficult. No OECD country has a pure market economy for innovation, especially when it has an important military sector. At the same time, all is not in the hands of the State, as free choices made by economic actors – in a framework of competition and optimisation – determine to a significant extent the innovation process.

In Russia, such a process is, at present, in the earliest stages of formation. In particular, an effective innovation system requires a business infrastructure, something that does not exist in Russia today. In addition, although the views and demands of society have changed dramatically and radically, structures are changing very slowly. To some extent, the previous system continues to operate, but in some respects it has collapsed: the old "rules of the game" have disappeared, but new ones have yet to be established. Thus, no one can yet predict what the future will bring.

In the past, the goal of the enormous science sector was not production or marketing but the fulfilment of the needs of the state planning system, needs which were motivated by prestige, were bureaucratically determined, and were more concerned with quantity than quality. Decisions to "innovate" and to introduce "innovations" into the "market" were based on bureaucratic/administrative fiat. Innovations were either ordered or were implemented in a manner that did not directly concern the inventors, who had little to gain or lose by the implementation of their process or the sale of their product and therefore little propensity to innovate. As a result, innovation was generally slow and ineffective. The military and space sectors offer exceptions, but they had huge resources, (nearly) unlimited budgets, and no need to consider cost effectiveness. When the client was the army, the market was clear and certain, and the customer as well as the provider knew the "market" and its needs. More generally, the notion of "innovation" as it is understood in market economies, *i.e.* encompassing the broad diffusion of new technologies, was unknown in the planned economy. There, as the Background Report underlines, a new technology, *i.e.* a new process or product, was operated or produced in only one or two enterprises (which enjoyed a monopolistic position by virtue of the system).

To accomplish the "task of innovation" in the state planning system, structures such as "science/production associations" or "inter-sectoral industrial complexes" were created in the 1980s. In a sense, these structures were intended to compensate for the fragmentation and compartmentalisation of the "innovation system". However, it is doubtful whether they will survive, and most of the large concerns that constitute Russian industry are in any case currently incapable of innovation. The Government has proposed the creation of financial-industrial groups to operate in the high-technology sectors, and this may provide new vehicles for change. However, establishing such large structures in the unfamiliar context of an emerging market-based economy open to foreign competition will be a lengthy and difficult process. As a result, it is crucial to develop an active sector of new firms whose activities are based on technology.

Two positive developments suggest the first beginnings of spontaneous innovation. First, there has been a rapid expansion in the number of science-based inventions. Moreover, inventors are motivated to innovate, because they see in innovation a means of earning money, although they do not yet clearly see how to achieve that goal. While this is not yet market-driven development, it is nonetheless a first tentative step. In the present circumstances, the first missing link is the lack of Russian entrepreneurs able and willing to co-operate with inventors. As a result, Russian inventors seek partners abroad. They are also motivated to do so by the current uncertainty: they feel that if Soviet/Communist power should return, it is safer to have the intellectual property rights abroad and therefore outside state control. In seeking partners abroad, inventors also often hope to be able to avoid taxation in Russia.

The second positive development – a smaller source of potential innovations than science-based ideas but possibly more fruitful in the short term – is the mushrooming of initiatives in factories formerly involved in relatively sophisticated production for the military sector but whose production in that area has dropped sharply. These factories often have very advanced tools and equipment as well as a skilled work force, and they are attempting to reorient their facilities towards civilian markets, even foreign ones when they can reach them. The "new" products sometimes are of a far lower technological level; for example, firms involved in building space stations or missiles have diversified into production of bicycles for export to eastern Europe. The problems encountered in these ventures are not tied to technical competence or a lack of entrepreneurship but rather to deficiencies of management skills and skills related to marketing, packaging, designing, after-sales services, etc. Here again, a foreign partner will be often sought for help in acquiring those skills or ensuring more stable production through subcontracting.

A large number of Russians, including scientific and industrial leaders, think that they have a "gold mine" of technologies that they can simply sell to the "West" for exploitation and marketing. The reality of the situation is quite different. Systematic screening of proposed technologies by specialised foreign agencies shows that no more than one technology out of a hundred is a serious candidate for commercialisation in an advanced market economy. This apparently low proportion should not be considered a particularly poor result, as it compares favourably with the "success" rate in a competitive world. The same failure rate is observed, for example, for inventions from US universities screened by venture capitalists.

A basic problem for creating a market-based innovation system is the attitude, widespread among the Russian population, that business is simply speculation. The Government must encourage a change of perspective so that normal business will be perceived as a respected profession, as current attitudes create serious obstacles to establishing more business-oriented structures. Regions of Russia differ with respect to views on private small business: certain areas which still have leaders from the previous regime have refused almost all private companies, while others, such as Nizhni-Novgorod, are very liberal and actively support business. It should be added that at the central level the importance of entrepreneurs is generally understood, but there remains a deep social and cultural problem which must be addressed.

The current upheaval has also generated various power struggles, in the course of which a battle has developed in the area of law making. Laws are enacted by various instances: President, Parliament and local "parliaments" pass often conflicting laws, thereby generating, at best, uncertainties as to the degree of protection or support that is accorded to innovators or entrepreneurs.

2. The intellectual property framework

Under the old system, the State was the owner of all intellectual work and thus of all patent rights. The inventor received a fee which was theoretically based on the economic value of the invention, but the fees were small. Many inventions were classified and were therefore not available for use in the open economy. International marketing of Soviet patents was centralised by a relatively inefficient state monopoly body, Lizensintorg; when a patent was marketed, the inventor received nothing. Moreover, lack of resources prevented the payment of fees necessary for maintaining patents taken out abroad. Thus, out of the 16 000 patents held in over 50 countries in 1990, 12 000 had lapsed by the beginning of 1993, according to the Background Report.

The main and revolutionary change in the climate for innovation has been the possibility for individuals – scientists and inventors – to be the legal owners of their own intellectual work and to hold the rights to their inventions. This gives them a personal interest in the commercialisation of their ideas. The value of this change did not go unnoticed. When it became evident that there would be very great reforms, scientists started to wait for "better times" and held back the publication of their inventions. This may have contributed to the apparently significant drop in the number of applications for patents and for "author's certificates", which decreased by 40 per cent in 1991 compared with 1988.

The new Patent Law of the Russian Federation was accepted by the Supreme Soviet on 23 September 1992. At the same time, the law on trademarks, service marks, and appellations of origin of goods was enacted. The Patent Law is written according to the European Patent Organisation's (APE) models, and it is the first step towards a modern "world standard" patent system in Russia.

However, it is only the first step, and it will be a long time before all the necessary bodies and procedures are fully operational and inventors are able to obtain the public support and commercial services they need. There is a clear need for careful organisation and for a substantial programme of training and dissemination of information which might benefit from aid from international technical programmes. The patent organisation, Rospatent, is in fact already in difficulty, because, as with many other state bodies, it does not have enough funds to accomplish its new tasks. Moreover, there are no enforcement mechanisms, no court of appeals, no apparatus of enquiry and control, and no penalty system.

Thus, even if there is a new Russian Patent Law and new legislation, scientists and inventors still feel sceptical. One result of the distrust of Rospatent has been that inventors prefer to obtain foreign patents rather than Russian ones. However, inventors and institutes do not have the foreign currency necessary to obtain patents outside Russia. For this reason, some have preferred to find a foreign company and obtain a joint patent; in this case, the foreign partner underwrites all the costs of patenting and is also responsible for the protection of the patent rights. However, Russian innovators also distrust foreign business, both because they are unfamiliar with business practices and ethics and because there are, inevitably, stories of abuse, in which Russians have lost the rights to their inventions. Access to better information about foreign co-operation is needed.

Doubts also arise concerning the protection of foreign intellectual property inside Russia, and there have not, as yet, been any instances of applications for Russian patents from foreign sources. Here again, the problem is the infancy of the new Russian legal system and the lack of good models for dealing with disputes over patent issues.

One outstanding question concerns what should be done with previously state-owned patents and who should have the ownership rights. The Baltic republics have established a procedure whereby old Soviet patents, originally taken out in the name of local institutes, can now be registered as new patents in the name of the original inventor. One solution might be to give patent rights to the institutes where the patent originated, as this would make property available to them for commercialisation in order to obtain income.

3. Support for innovators and entrepreneurs

As previously noted, institutes specialised in applied sciences are an important source of potential innovation; they have had closer connections with industry and thus some experience with innovation. However, now that industrialists are experiencing all the effects of the economic crisis, they have lost nearly all interest in financing R&D outside their own factories, so that even the more innovation-oriented applied science institutes have seen their positions further weakened by the simultaneous loss of their clients and their financiers.

Individual scientists, on the other hand, have had very little experience of innovation and even less of the risks of the business world. Today, many keep their jobs in the state sector but also work part-time in the private sector. They are generally not ready to take the risk of becoming entrepreneurs or starting their own businesses, both because they do not wish to lose the security offered by their state positions and because of general uncertainty about the future. (For the same reasons, those who leave the country to work abroad try to maintain their institute positions at home.) In general, the more business-minded scientists have left the science sector completely.

Technical, commercial and financial assistance

A glaring deficiency in the present situation is the lack of support mechanisms for individual innovators and entrepreneurs. One very serious result of the lack of available hard currency is that greatly needed foreign information – in the form of journals, books, and reports – is not available. Scientists and engineers often cannot use foreign computer networks to get information because of the high price of communications, including telephones. Some new academies and state funds are trying to create such support, but they require foreign technical aid for literature (and translation), journals, and computer networks, as well as for national information networks.

Most entrepreneurs know nothing of export markets – of market needs, technical or safety standards, user manuals, official controls (registration), prices, packaging, etc. – and will be unable to do without foreign participation in marketing. There is good export potential for firms in the high-tech science industry, but if this potential is to be developed, new structures which link such companies to world markets and remove obstacles to export – in the area of laws, taxation, licences, fees, etc. – must be quickly put in place. The rules for export must obviously be stable and perceived as being so by both exporters and importers.

While some state funds are available for financing commercial efforts, they are more oriented towards organisations than towards individual entrepreneurs. Existing funds resemble ministry subsidies rather than profit-oriented or venture-type funding. Funds can be a good and flexible tool for promoting innovation, and, if they are local and independent, they can help boost development and commercialisation. Foreign participation in the financing of funds offers a form of commercial co-operation, and foreigners can play an important role in evaluating potential innovations. Funds with foreign participation can help ensure that innovations reach world markets and help protect the rights of the innovating Russian enterprise. The banking system is now undergoing comprehensive reform, and it is difficult to assess how much support it can provide new entrepreneurs. At present, government funds are managed by informal networks, and real interest rates are extremely low due to the inflation rate. As a result, an entrepreneur, providing he is well connected to these "networks" and has a promising project, encounters little difficulty in finding funding, particularly since no risk evaluation is involved.

Institutes need financing, they are overstaffed, and they have too many physical facilities. One small-scale and effective way to help inventors, and institutes as well, is to rent to them, at a reasonable fee, unused institute laboratories, so that they have the laboratory capacity, infrastructure, subcontracting services, and other means they need to start a small company. This is an alternative to more ambitious plans to develop technology parks; it keeps the fledgeling businesses closer to the institute and gives the scientists a view of the market. This is a good way to ensure better communication between scientists and entrepreneurs and should be more widely practised. Some attempts have been made in this direction, but more official support for such initiatives is needed. A combination of local and central support for "science business" also helps science financing.

Minnauka has a special programme for science parks and funds to start five "model" parks in 1993. In principle, this programme should foster science-led innovation. However, science parks have generally been treated as real estate projects, rather than as programmes to help inventor-entrepreneurs create new industry. As now understood, science parks are unlikely to help generate new science-based businesses. In simply practical terms, potential entrepreneurs find it difficult to get office space, telephone, fax, marketing, components, packaging, and, not least, information, and this is what innovators and entrepreneurs need in priority.

Entrepreneurship in manufacturing

There are a great many new Russian entrepreneurs, but the majority are in commerce, selling and buying consumer goods for the population. Entrepreneurs who wish to start their own production business face far greater difficulties than those in trade. The level of taxation places a serious burden on new manufacturing businesses. The taxes levied on manufacturing are not greater than those on services, but they are more onerous, given that in market activities gains can be much more rapid and much higher. Tax incentives can be an important policy tool, but it will not be possible to use them until the situation of the entire tax system has been clarified and stabilised.

Entrepreneurs ask for state support, and there is a long tradition in Russia of state/city participation in businesses. Private enterprises and new entrepreneurs are very much dependent on state or local government organisations. Because of the high degree of government involvement – in terms of raw materials, logistics, production plants, export licences, etc. – private companies function only partly under market economy conditions. Local monopolies can ask higher prices than they would receive on world markets, and the price level will only become normal (in international terms) when there are (free) imports. Monopolies have to be controlled, and taxation is a strong means of control.

Moreover, industrial infrastructure (machinery, buildings, standards) is made for large firms, not for small companies. As there are no programmes to split up the former – because of "internal integration" – the latter will have to start from scratch, with pilot plants, etc. Few firms will break off from big enterprises. In such a context, anti-monopoly laws – which have begun to be enacted but do not have adequate enforcement mechanisms – become crucial. More fundamentally, there still exist in many parts of Russia "informal" groups which impede trade and the transport of goods, practices which need to be curtailed by police intervention.

4. The involvement of foreign industry

It is clear from what precedes that foreign industrial involvement is likely to be a very important source of dynamism and change for the exploitation of Russia's innovative potential.

Subcontracting by industry in OECD countries to Russian scientific institutes is a type of co-operation that can benefit both parties, thanks to the high level of know-how and scientific expertise and the low salary levels in Russia. The mechanisms for such arrangements are not yet clear. At present, many models are being used, ranging from personal oral contracts to official agreements between institute management and the foreign partner. It would be useful to have an inventory of the different formulas now being employed.

Also, technology transfer from Russia to OECD countries is one way of commercialising the results of Russian science. Here, it is important to protect the interests of inexperienced Russian scientists and engineers when they engage in legal agreements, for instance by establishing standard, internationally agreed contracts. On their side, the Russians should not sell the same thing to several parties on an "exclusive" basis.

The principal problems encountered by OECD country firms have already been mentioned. They include: transferring funds and payments to Russia, obtaining visas and going through customs, telephone communications, problems of language, and the poor quality of life which makes it difficult to send teams of researchers and engineers for long periods of time. To these must be added a certain number of specific problems linked to industrial and commercial goals: the absence of effective patent protection; the impossibility of obtaining locally techniques, instruments, and software that are readily available in OECD countries; and the obstacles created by the COCOM system, even though many restrictions have been removed, and Russia's own restrictions on technology exports.

It is generally felt that the efficiency of R&D co-operation increases when it takes place in the framework of plans to undertake manufacturing production in Russia.

For collaborative ventures in raw materials production, the Russian partner is a state or state-controlled industry. Russian decisions are often made by government and are political in nature; these ventures are therefore not open to competition among firms. In most cases, foreign technology is needed, and this gives an immediate return on foreign investments. Investors and foreign partners also receive income from the export of raw materials or energy. In general, the foreign companies act as subcontractors and technology suppliers, and the business remains in the hands of Russian industry. Foreign companies can participate by making alliances with Russian industry and by making offers to participate from the beginning.

Foreign industry can also participate in Russian industry by buying shares in existing industrial capacity and establishing joint ventures. Shares are sold to foreigners in order to obtain new technology or a new marketing partner and because foreigners can pay higher prices. The first wave of privatisation has been organised by the State, and the new owner can later sell to interested foreign companies. Foreign firms participate in order to gain access to Russian markets and to produce more cheaply. It is also a way for medium-sized companies to become established in Russia. Joint ventures, which developed rapidly in the early 1990s, now have almost ceased, partly because the general conditions of privatisation remain unclear; tax advantages and facilities for repatriating profits have, moreover, been seriously reduced. Such co-operation is likely to grow rapidly when the internal situation in Russia becomes more stable. However, this kind of partnership encounters considerable cultural difficulties, and each side needs to be educated about the other.

Even more than for R&D, subcontracting is undoubtedly the best form of co-operation, because the economic risk for the foreign partner is low and because Russia can offer subcontracting services at competitive prices. Russian subcontracting partners possess factories, production capacity, good workers, and raw materials. Assembly of products containing foreign or Russian components, or both, is another good form of co-operation, as it does not require large investments in production and can take advantage of inexpensive skilled labour. In order to attract foreign firms to subcontract, the Russian Government can offer tax incentives, such as a tax-free period and lower currency taxation, and other simple forms of economic support.

5. Normalisation of technology: standards, certification, quality control

Standardisation and quality control are important issues for restructuring. In light of the very limited resources and the overall economic situation, consumer protection does not yet have very high priority, although new trends can be seen in the attention given to international standards and European practices. Although problems of quality clearly exist, they tend to take second place to issues of quantity, for reasons that have partly to do with past attitudes and traditions. All actions related to normalisation of technology are fundamental, however, as they are the first prerequisites for building a domestic market that can be gradually integrated into the global economy.

It should be noted that the former Soviet Union had developed a remarkable infrastructure for standardisation and normalisation of technology. The Russian State Committee for Standards has inherited a large part of it, including a staff of more than 30 000 persons, more than 100 regional offices, and about 20 research institutes, but the lack of resources prevents making good use of it, and this infrastructure is deteriorating quickly.

The Soviet Union participated in the work of the International Standards Organisation (ISO) but had its own system of state standards, the GOST system, which was used in all Comecon countries. Today, the Baltic countries, as well as Hungary, Poland and the Czech Republic are leaving the GOST system and have become members or associate members of the Conseil européen des normes (European Council on Standards – CEN) and ISO. The PHARE programme of the European Community (EC) and the European Free Trade Association (EFTA) help these countries as they adopt European standards, and Sweden gives economic aid to Estonia for standardisation. Because of the long-standing GOST tradition, this is proving to be a very slow process.

Russia has entered into agreements with some NIS countries concerning participation in the international standardisation work of ISO. Russia will act as the official representative of these countries in ISO and in other bodies involved in international standardisation but finds it difficult to pay the fees for participating in ISO. Russia has also shown interest in CEN, but so far the EC has not responded, possibly for political reasons.

Russia's system of standards is GOST-R, which, because of greatly reduced funding, finds itself in difficulties and has become far less effective. The philosophy behind standardisation was and is different in the former Soviet Union and Russia from what it is in the developed market economies. For example, most specific consumer products, such as bread or vodka, were and are the same throughout the country. (In western Europe, instead, such product standards are in the process of disappearing.) Russia is now moving towards the use of certification for consumer protection. This is a difficult process and only 30 per cent of goods have compulsory certificates today. CEN standards are the basis for certification, but there are some departures from them.

There is also the fact that Soviet industry was unaccustomed to evaluating quality, as its production was measured by quantity, the criterion imposed by the GOSPLAN. The lack of real markets meant that consumers did not complain about poor quality, accepted whatever products were available, and were happy to queue for them. After-sales service was also badly organised. However, some Russian factories now have ISO 9000 certification, and as they export to mature market economies, the level of the quality will rise.

Standardisation, certification and quality control are all important aspects of the industrial infrastructure and can be used to help Russia integrate into world markets.

6. Concluding remarks

It is clear that it will take time to build an appropriate innovation system. Infrastructure, experience, and motivation are badly lacking. It is important to establish a judicial power with the capacity to act in the areas of science, technology and industry. This is essential for enforcing the application of laws protecting inventors (patents), entrepreneurs (competition), and consumers (regulations), who must have the possibility of recourse and the means to assert their rights.

Privatisation will promote innovation, and widespread privatisation can accelerate the creation of the infrastructure, markets, and financing systems needed for a more market-based innovation system. It also changes attitudes and gives motivation for starting businesses. The stabilisation of the economy and the cooling of inflation are also indispensable, since without stabilisation, it is impossible to run a normal business.

Foreign governments could also do more, in the form of bilateral and multilateral aid, to help improve the innovation climate. Foreign interest tends to concentrate on the science base or sensitive technologies but is inclined to neglect fundamental areas such as the infrastructure for patents, standards, etc. Moreover, only a few foreign countries seem to pay real attention to support to small and medium-sized firms when developing collaborative schemes.

In an economy without previous experience of such activities, the role to be played by small firms in Russia remains an open question. The individual efforts of entrepreneurs will doubtless have to receive some form of community support, from the regions, from associations, and also from the large firms with which the small ones will necessarily coexist.

Chapter V

From military to civil applications

Having taken up in the preceding chapters the broad lines of a research and innovation policy that seems appropriate to the transition period, this chapter is devoted to the strategic question of reorienting the overall development process by reducing the share of science and technology invested in military applications and increasing considerably the share allotted to civil applications. This is not simply a question of budgetary transfers and reorientations. Indeed, what is involved are two fundamentally different ways of looking at things, and the concerned actors have as yet very little familiarity with the conditions of civil S&T. The discussion begins with the issue of reducing military S&T, including the implications for redirecting the activities of the closed cities, which were entirely focused on military goals. Next, the reorientation of the space and aeronautic sectors towards civil goals will be evoked. Requirements for developing S&T in other sectors, in particular those involved in natural resources (energy and agriculture), will be sketched out. A discussion of the modernisation of the telecommunications network, an essential investment for the entire development process, follows.

1. Military S&T

Throughout the post-war years, Soviet R&D was heavily militarised, with unfortunate consequences for civil S&T capability. This hypertrophied military S&T system was concentrated in the Russian Federation, which now has on its territory 85 per cent of the R&D personnel of the defence industry of the former Soviet Union, a sector which, in terms of expenditure, was responsible for almost

90 per cent of former Soviet military research. The Russian "defence complex" (*i.e.* those enterprises and organisations overseen by the Committee for the Defence Branches of Industry and the Russian Ministry of Atomic Energy) has some 700 research and design bureaus, which employed, at the beginning of 1992, approximately 1.3 million people. In recent years these defence complex S&T organisations have received some 70 per cent of state budget allocations to R&D. In addition, military-related R&D has been undertaken by institutes of the Academy of Sciences and also by some of Russia's elite higher education establishments.

The former Soviet defence industry always had a role in civil production, and it was also an important performer of civil research: in recent years, approximately half of all civil R&D supported by the state budget was undertaken by organisations in the defence complex. This civil S&T activity includes most of Russia's research into civil aviation, shipbuilding, electronics, information technology, advanced materials, civil space technology, and the development of certain types of advanced production equipment.

The research institutes and design bureaus of the military sector tend to be relatively large, often employing several thousand people, and, compared to their civil counterparts, they have relatively good equipment. A large proportion have been incorporated into science/production associations with facilities for small-scale production. Until recently their personnel worked under conditions of extraordinary secrecy. This secrecy, coupled with the usual administrative logic of the former Soviet R&D system – possibly even more strongly expressed in the military sector – led to considerable duplication of facilities and research effort. Secrecy also protected research personnel from peer group assessment: academic titles and awards were obtained without external review. Pay and social amenities for military sector R&D personnel were superior to those available to most of those who worked in civil S&T. These conditions undoubtedly fostered some genuine research excellence, but they also served to protect inferior work and allowed people without real talent to obtain unjustified status in the scientific community.

Military R&D is experiencing acute difficulties created by serious budget reductions (see Chapter II, Section 3). To make matters worse, non-budget sources of finance have almost disappeared. This means that science and technology in the defence industry are becoming increasingly dependent on military work at a time when "conversion" is the declared priority. It has been claimed that, in 1992, 200 000 people left the defence sector S&T system, pay levels collapsed, and new recruitment virtually ceased.

The Russian armed forces are attempting to elaborate a new military-technical policy, and the sooner this is done the easier it will be to devise appropriate policies for defence sector S&T. Present indications are that they give considerable priority to dual-use technologies and recognise the need to underpin military R&D with a dynamic civil high technology research base. The industrial policy proposals emerging from the Ministry of Defence offer potentially promising solutions. The creation of a set of diversified high-technology corporations engaged in both civil and military work, with private, state, or mixed forms of ownership, would create conditions for the survival of some of the country's strongest branch S&T teams. Those R&D organisations not vital to the realisation of the new military-technical strategy would be free to civilianise, provided that they can find ways to meet civil demands. Many defence sector R&D organisations have broad, diversified profiles, and this offers possibilities for separating out viable departments and converting them into independent units. However, it is likely that many defence sector S&T organisations will not be able to adapt, making closure the only viable option.

2. The problem of the closed cities

Military R&D is concentrated to an extraordinary degree in and around Moscow, where approximately 350 institutes and design organisations are located and are responsible for about half of Russian R&D in aviation, 40 per cent in electronics, and two-thirds in the radio industry. The geography of military S&T presents another severe problem: the phenomenon of "closed" cities dedicated almost exclusively to particular lines of military R&D. The ten closed cities of the nuclear sector are now well known, but there are up to 70 more in other fields, including 16 in the Moscow region, *e.g.* Kaliningrad (missiles and space technology), Zelenograd (electronics), Zhukovskii (aviation), and Fryazino (electronics). Not only do these cities have little or no alternative employment, but most of their housing and social welfare facilities have been funded from ministerial sources.

Restructuring presents the closed, or formerly closed, cities with an extremely difficult situation. By the nature of their activities, some – especially those concerned directly with the assembly or dismantling of nuclear devices – will retain some security restrictions, but others should be opened up as much as possible so as to permit their full integration into the economy and the involvement of foreign capital. In order to solve the problems of these mono-activity towns, it is vitally necessary to transfer their social infrastructure to the local authorities, and to accomplish this, a new local taxation system is essential, as is the privatisation of the housing stock. This is a more general problem and extends beyond the closed towns: its solution will enhance labour mobility and ease the social costs of unemployment. Some of the closed towns could become centres of small high-technology business, possibly with foreign backing in the early years, although experience in the OECD countries suggests the need for a sober assessment of the likely economic benefits to be derived from the creation of technopoles, science parks and similar concentrations of S&T activity, especially if they are formed in the absence of clearly expressed market demand. As the local economies of the "dedicated" cities diversify, it is likely that some of the research and engineering personnel will find employment outside the S&T sphere.

3. Space and aviation: the shift from a military to a civil focus

Space and aviation are sectors that were strongly developed in the former Soviet Union and are currently engaged in a gradual shift from a military to a civil focus. In aviation, Russia has considerable R&D potential; under the Soviet Union, the main accent was on military applications, but a civil aviation sector was also developed and consolidated under Aeroflot. The problems involved in further developing civil aviation today are complex. Aeroflot has fragmented into more than 100 local companies. There are insufficient funds for maintaining aircraft and ground facilities, and certain sources report that there were some 20 accidents on the territory of the NIS in the course of 1992. Facilities for training pilots and ground personnel to meet international standards are sorely lacking (a project to build a training centre with a flight simulator at Zhukovskii, initiated by the OECD with the backing of the Canadian Government, at a cost of $130 million, is presently under study).

In the area of research, development, and innovation, strictly speaking, the problems are due not only to the lack of funds but also to structural inadequacies. The separation of research centres, design bureaus, experimentation sites, and production factories does not facilitate the innovation process, as each element tends to develop its strategy independently. Furthermore, revenue earned from production is not used to support research and design, even though all belong to the same group (Mig, for example). Several important agreements – at least from the point of view of the Russians – have been signed with foreign aviation companies: to adapt western engines and cockpits to Russian airframes, to exploit sophisticated calculation software developed by the Russians, to produce aircraft parts through subcontracting, etc. According to certain estimates, the prospects of progressive integration into the world market are not negligible if a coherent and consistent strategy can be defined. In an initial phase, Russia would certainly be able to satisfy the markets of developing countries, and later – in ten years or so – a proven track record would open markets in the more advanced countries.

The space sector, in which the Soviet Union invested massively (more than 1 per cent of GNP in 1989) employed more than one million persons, of whom 600 000 were in research and development. More than 90 per cent are now in Russia, and 40 per cent of these are in the Moscow area. Recent reductions in the government budget have been substantial, since, in real terms, the budget for 1992 is only half that of 1991. Russia has thus been led to step up efforts to co-operate with foreign countries by renting or selling its services, including its satellite-launching capabilities.

Russian strengths are many: for long-term space flights, they are well ahead of the United States and Europe in terms of overall experience and research on human factors. Russia has launched or planned many missions beyond the Earth's orbit, and this has made it possible to collaborate with foreign partners in missions towards Mars and other planets. Scientific co-operation is developing with NASA, in the area of space flights and long-distance missions, and with the European Space Agency (ESA), notably for the follow-up of the Hermes space vehicle. Present budgets are of the order of $100 million in both cases for programmes planned over several years. As in scientific co-operation in other domains, the contracts are largely signed with institutes and research teams. It is important that the newly created space agency (RKA) adapt its practices and reinforce its co-ordination mechanisms, in order to gain credibility both with foreign operators and with space-related industry in Russia itself.

The Proton launch rockets and others are among the most powerful available. The vast stockpile of ICBMs offers further launch capabilities. The prices offered by Russia for launches are considered in competing quarters as close to dumping, since Russia has recently offered, for example, prices at a quarter or a third of those of Western launch rockets, which are of the order of $100 million. The dumping tendency undoubtedly exists, but it should be recognised that Russia is not presently able to determine precisely the real costs of its products. Agreements to set the price of Russian launches within an acceptable "window" are under study with the United States and Europe; these agreements would also accord the Russians a certain quota of launches up to the year 2000. Until the skills and data necessary for effective costing are available, dumping claims are likely to continue.

It is important to recognise that, be it in aviation or in space, Russia should try to price at real prices, despite the difficulties involved. Otherwise, there is a substantial risk of seriously disturbing world markets, and negative effects on OECD country economies will rapidly have significant repercussions in Russia as well. As the renewal of Russian industry and its capacity to reach levels that allow it to compete in world markets presently depends on co-operation with counterparts in the OECD countries, it is necessary to seek "co-operative equilibria" that will suit the needs of the different partners. These observations also apply to other areas, such as non-ferrous materials, in which tendencies to create a worrisome imbalance in world markets can be seen emerging.

4. Developing civil S&T

Many sectors of the Russian civil economy have a relatively backward level of technology when compared to that of leading OECD nations. The uncompetitive character of much of the country's industrial base makes it difficult to expand exports of manufactured goods to demanding foreign customers. The energy sector, the principal source of export earnings, requires extensive modernisation. Domestically produced consumer goods fail to match market economy standards of design and quality. Agricultural technology developed to meet the perceived needs of large-scale state and collective farms and neglected the needs of smaller-scale farming. Chronically underdeveloped telecommunications hinder economic renewal, and some technologies vital to the creation of a modern service sector are almost entirely absent. In many sectors of the economy, the inherited technological base is damaging to the environment, wasteful of resources, and hazardous to the work force and the population at large. There is no dispute that Russia needs new civil technology on a substantial scale in many sectors and that this represents one of the central strategic issues now facing Russian science and technology. The problem is to meet this need under conditions of budget stringency, at a time when market institutions are in the process of formation and foreign investment is inhibited by domestic uncertainties.

Given the acute budget constraints, government funding for civil S&T is likely to be severely limited for the foreseeable future. This means that state support will have to be highly selective and that alternative non-state sources of finance must be promoted. In pursuing a selection policy, the Government will need to be resolute in withstanding pressures from powerful bureaucratic interests. It is desirable that civil S&T should, to an increasing extent, develop in response to market demand. Initiatives to foster demand-led civil S&T activity are of vital importance, given that the supply side almost totally dominated the former Soviet economy. However, in circumstances of economic decline, high inflation rates, and acute payment difficulties for enterprises, customer demand is now weak or non-existent in many sectors. This also applies to consumer markets. Unfortunately for the domestic economy, the new business class, which does have spending power, has a strong propensity to buy imported luxury goods. As some enterprises undergoing conversion are finding, the manufacture of such goods under licence may offer a profitable means of entry into new consumer markets. Demand-led technology development may also be possible in the agricultural sector; for example, the development of new farm equipment could be stimulated by the creation of a network of rural development banks charged with advancing loans for the purchase of machinery to the emergent private sector. In other sectors, the development of civil S&T could be promoted by the adoption of new procedures for government procurement designed to stimulate the supply of cost-effective, competitive technology.

The modernisation of the health care system – often considered dilapidated by observers from OECD countries – would seem to offer good possibilities for priority action in the areas of innovation and R&D, via standards and procurement policies. In fact, it would be important to know whether it is possible to coordinate the activities of penniless hospitals, which seem today to be answerable to many authorities (ministries, localities, firms, etc.). In addition, the maintaining and reorganisation of a system of social welfare is another important element of the modernisation of the health care sector. This is, clearly, a domain with significant possibilities for technology transfer from the military sector for medical instruments, radiology, nuclear medicine, etc.

Energy is a priority sector of Russian civil S&T. There is already active foreign participation, and, as part of conversion, domestic defence industry R&D organisations and enterprises are also involved. These developments are encouraging, and there is evident scope to expand activity, although the propagation within Russia of exaggerated fears of foreign dominance of this strategically important industry work against foreign contribution to the modernisation of the technology of the energy sector. Russian S&T development would be more effectively promoted if foreign companies involved in this and other sectors paid more attention to making use of domestic skills and production capacities. Not all foreign firms have realised that greater sensitivity to Russian national feeling may pay dividends from a business point of view.

In energy, considerable progress of great economic value could be realised by relatively minor maintenance and repair efforts requiring little in the way of technical innovation. Leakage of oil at the well site and along pipelines appears to exceed one million tonnes per day. The gas used in transfer and pumping operations appears to exceed 30 per cent of the total amount transported (against 5 per cent in the OECD countries). However, as long as the price of energy remains several times below international market prices, there will be little incentive to economise. The same argument applies to the not inconsiderable economies of energy that could be achieved in domestic use, transport, etc.

Russia's acute environmental problems are universally recognised, but domestic sources of finance for solving them at the national, local, or firm levels are likely to remain severely constrained for many years. This is a prime area for foreign assistance. Insofar as new technological solutions are required, the Russian defence industry is in a good position to provide them. There is clearly scope for more active international involvement in a conversion programme that tackles the environmental situation that Russia has inherited. More generally, modernisation of the Russian economy should take into account the environmental constraints now imposed on the richer countries that are helping advance the modernisation process. In a framework of co-operative equilibria, Russia should also make systematic efforts to improve the level of environmental impact throughout the industrial sector.

5. Telecommunications: a key investment

There is a high correlation between level of development and number of telephones, and the cause-effect relationship is from the latter to the former. This is because the telephone is the basic instrument of business, as it makes possible the free and flexible development of projects and thus augments overall economic activity. The former Soviet Union – of which Russia was the midpoint – has about 15 telephone lines per 100 inhabitants (according to recent OECD and International Union of Telecommunications statistics) against some 44 for western Europe and 51 for the United States. Current waiting lists for installing connections correspond to just under half of the existing network, but this may underestimate suppressed demand. The quality of the public network is inadequate and highly variable; this make data communications, including fax transmission, very difficult. Digital switching and fibre optics cables are only used to a limited extent, mainly in former military and some scientific networks. The current proliferation of mobile telephones is a useful stopgap measure, but it has certain limits (saturation of wave lengths and access restricted to privileged groups). In the age of information, communication difficulties represent a basic impediment to Russia's integration into the international economic community. Foreign investment will naturally go to places where communication with parent firms is assured, and there is intense global competition for the limited investment capital available. In a recent OECD survey, enterprise managers in six central and east European countries, including Russia, found the low quality of telecommunications to be the main domestic infrastructural barrier to foreign trade. Investment in this basic infrastructure is urgently needed, and it will require help from international organisations and financing.

The lack of strategic planning for the telecommunications network is a major difficulty. If the Government lacks the funds to put a communications system in place, commercial interests will do so. In the 19th century, a comparable situation existed for railroads: in an attempt to create a rail network that satisfied both military and defence requirements and met the demands of commerce, hesitant government policy succeeding in satisfying neither. This suggests that the Government must take an active role in arbitrating competing interests and in helping to direct limited resources from both public and private sources towards an overall framework. It is not that the State should take total control; such an approach failed in the past and would only drive out private capital. Rather, the Government must help co-ordinate and facilitate the efforts of domestic commercial interests, foreign investors, and donors, while providing some state support. The Ministry of Communications should consider separating operation from regulation, for example by allowing the operators to be incorporated and allowing tariffs to be set according to costs. A separate but important contribution to rapid network development would entail the conversion of the overall communication resources of the military/industrial sector to civil use.

As far as the transmission of data is concerned, the Russian Federation is mainly served by the packet-switched RELCOM electronic-mail system (with 90 per cent of the market). The system reaches at least 60 000 subscribers, but is outdated (UCCP Accord). The primary obstacle to a rapid rise in the power of interactive services in Russia is the poverty of high-speed digital lines (64 Kbps at most). Although new business-oriented suppliers are emerging, the relatively high prices of their services may put them out of reach of most scientific institutions. In this context, the Ministry of Communications and the Committee for Informatisation have a special responsibility to make available high quality data communication facilities at prices affordable to the scientific community.

6. Concluding comments

The shift towards a development strategy founded on the civil rather than the military sector requires a series of quite momentous changes in attitudes and in the institutions involved.

First is a need to subscribe deliberately to a strategy of détente and peace and to understand that, today, the future and the power of a nation depend on economic rather than on military power, even if a certain capacity to export arms can help ensure wealth and maintain employment.

Next, it is necessary to see the State as being at the service of the well-being of the population, in terms of health care, transportation, the environment, etc. It is also important to develop a sense of public service and fair and competitive practices for public projects, attitudes which seem, paradoxically, to have been entirely neglected under the Communist regime, despite its strong community tendencies and the power of the state structure.

Last, and far from least, it is necessary to stop thinking solely in terms of technologies situated at the forefront of current knowledge and capabilities and to put in place a policy of accommodating and distributing less sophisticated but well mastered technologies that answer the needs of the market. While they may wish to be a major power in terms of world S&T, Russian policy makers might reflect on the post-war experience of Japan. Japan did not begin with high-technology products, but rather played at the start on their comparative advantage, a highly skilled and relatively inexpensive labour force. Russian policy makers should not be so caught up in prestige and parity that they sacrifice what may be their greatest short-term advantages.

In addition, "lower" technology is precisely what is needed in the Russian domestic economy. Satellites, space stations, and superconducting supercolliders will do little to aid agriculture, food processing, transport, or other aspects of the consumer sector. Choices regarding resource allocation from the limited state budgets for these purposes will clearly be made by planners and politicians. They should make their choices on the basis of the issues involved, and they should neither try to satisfy constituencies advocating the continuation of obsolete programmes nor sacrifice to the persistent myth of technological avant-gardism.

Once these basic observations concerning attitudes and mentalities have been made, the question of the orderly reduction of military S&T and the deployment of civil S&T in Russia remains open. In truth, the outlook in the short and medium term appears rather uncertain. To what extent, in fact, can the public authorities weigh effectively and in a co-ordinated manner on the different actors concerned by the modernisation and the reorientation of the technological effort? In addition to launching appropriate technology programmes, they will have to implement effective reforms in many areas if they are to resituate science and technology in society and the economy: fiscal regulations, social services, local responsibilities, foreign investment. The decision process seems now to depend on innumerable individual interests and bureaucratic powers. A combination of the preferences of these actors may not offer the best solution for the society as a whole.

Finally, the magnitude of the variables at play and the sensitivity of the key parameters involved can have important industrial and commercial consequences. For this reason, there is a pressing need for Russia and its foreign partners to elaborate a framework for creating the "co-operative equilibria" necessary to an adequate balance between Russian interests and those of OECD countries. The effort to establish appropriate "rules of the game" in this area is an important step in facilitating the progressive integration of Russian science and technology into the world economy.

Chapter VI
Principal recommendations*

Science and technology were one of the prize jewels of a system now undergoing profound change. It grew according to a kind of bureaucratic logic and in response to military needs and became very large, although its real cost to the nation was difficult to judge. Today, in a situation of crisis in which science and technology now have low priority, the country can no longer maintain an organisation of the size and type that it has inherited. In fact, the erosion of Russian S&T is well under way, primarily because of the lack of financial means.

The authorities are attempting to avoid excessive confusion in this period of change, and, overall, the measures that have been taken seem to go in the right direction. However, the problems are immense, and there is resistance to reform within the communities concerned. Further, given present conditions, the actions that government authorities can take with respect to science and technology have certain limits.

In this concluding chapter, we shall try to draw together the principal recommendations sketched out throughout the document. Given the rapid evolution of the situation in Russia, it should be remembered that these recommendations were drawn up in the summer and early autumn of 1993. In what follows, we shall restrict ourselves to what seems essential; more detailed information can be found in the body of the report. We shall follow the major themes addressed in the various chapters: institutions, research, innovation, and civil technology. Special attention will be given to the internationalisation of science and technology. Finally, we will allude to statistical and accounting problems.

---

* See the summary record of the September 1993 review meeting in Moscow for observations made by the participants and for complementary information.

1. Reorganising the scientific and technological policy institutions

As mentioned, the measures already taken in this area are not insignificant and seem to go in the right direction. Points on which we feel that it is necessary to hold firm or advance are the following, at successive levels of the State:

i) **The creation at the highest level of government of adequate co-ordination mechanisms through which the place to be assigned to science and technology as part of putting the country on the road to recovery can be debated and defined,** along with the extent of the financial effort that should be made and all the general economic or political measures that are appropriate for promoting reform in the area of research and innovation (regionalisation, taxation, etc.). A co-ordinating structure, under the chairmanship of the Minister of Science and Technological Policy, brings together representatives of all concerned ministries. Another body, with a higher status and with participation restricted to the ministers most closely involved (Science and Technological Policy, Defence, Finance, Industry, Education, etc.), might perhaps be set up, under the chairmanship of the Prime Minister, with the Minister of S&T policy assuring a key secretariat role.

ii) **Clarified allocation of responsibilities in the management of science and technology in the central government.** The responsibilities of the Ministry for Science and Technological Policy should be maintained, and, if necessary, strengthened, for the overall civil R&D budget and the broad lines of S&T policy. The Academy of Sciences should continue to manage its institutes and remain the "shrine" of Russian science. The authorities in charge of higher education should increasingly participate in the definition of research policy, and the universities' involvement in research should gradually broaden. The Russian Foundation for Fundamental Research should remain an independent agency, in the management of which the Academy and, in time, the universities should be closely involved, within the ambit of the ministry.

iii) **Development of regional competencies in S&T policy.** This is an essential point given the present evolution of Russia and the federal nature of the State, which should lead to growing decentralisation.

While it is important that the central institutions systematically establish regional agencies, a part of the national R&D budget should also be allocated to the regional authorities. These, in turn, should allocate to S&T a percentage of the increased resources that they are likely to collect by virtue of new constitutional arrangements, and they should, in consequence, develop their capacities for managing science and technology.

iv) **Exploiting local S&T capacities in the cities.** Russian cities with significant national or international R&D capacities should concentrate their efforts on these resources and support individual or collective initiatives designed to exploit them. The creation of science or technology parks is not necessarily the best solution; they should only be developed on a pilot basis. Rather, innovators and researchers can be supported by actions that are less costly but more immediately relevant to their needs. The science cities (naukograd), a number of which worked exclusively for the military, present particular problems and will encounter difficulties as they attempt to diversify. Institutes and enterprises need to be restructured, and to the extent that local financing is increased from local taxation, they should no longer be held responsible for financing local housing and social services.

v) **Decompartmentalising the research and innovation system.** The Russian research and innovation system was, and is, extraordinarily compartmentalised and hierarchical. As OECD country experience has made abundantly clear, innovation progresses by means of the cross-fertilisation of ideas. The breakdown of barriers and the increase of communication among S&T fields and sub-fields should be encouraged by all available means: mobility, diffusion of information, joint R&D programmes, common workshops and seminars, etc. It is essential that communication be facilitated between the Academy and the universities, and, to this end, researchers should be encouraged to teach in universities.

2. Adjusting resources invested in R&D to economic capacities

Many groups within the scientific community wish to see the country's R&D capacity maintained as a highest priority. However, R&D capacity was clearly too large in a number of areas (while insufficiently developed in others). Science fills an important role in society (and for humanity in general), but it is not a direct source of the creation of wealth. It is therefore necessary to determine, in a clearheaded way, what Russia, given its means, can afford to invest in science today. Under the pressure of circumstances, the scientific research structure is being reduced in a rather anarchic manner. Achieving the necessary reduction in a more orderly and rational way presupposes a certain number of conditions.

i) **Allocation of a reasonable R&D budget within the state budget.** The amount has now apparently dropped to 3 per cent, which seems "reasonable". Budget periods should not be shorter than one year, and research institutions must be informed of the amount of the budget and be assured that the funds will be forthcoming. Otherwise, it will be impossible to maintain a major R&D system.

ii) **Progressive reduction of the number of researchers and technicians.** The inclination of the Russians, admirable as it may be, to refuse layoffs and to distribute available resources equally does not promote the reduction of overall numbers nor does it help retain the best. However, it seems necessary to lower the clearly excessive number of persons employed in R&D, in light of present economic capacities. Given their qualifications, as compared to those of the rest of the population, much of the personnel concerned should find employment.

iii) **The development of rational reduction and reorganisation plans within the different R&D institutional structures.** This concerns the Academy of Sciences, the universities, and branch institutes that have been strongly affected by present circumstances. The Ministry of Science and Technological Policy's project to create federal research centres in order to save the most useful branch institutes seems to be an appropriate procedure, if it is executed in an appropriate way.

iv) **The development of peer review evaluation mechanisms for projects, researchers, and institutes.** To make the necessary choices in the most judicious way, peer evaluation procedures should be used. If Russians are not accustomed to this practice, they have the means to conduct it, and the recent experience of the Russian Foundation for Fundamental Research shows that they seem to be able to implement it.

These peer evaluation mechanisms, despite their failings, are preferable to arbitrary administrative decisions or an uncertain process that can eliminate talented research teams that are poorly integrated into the scientific establishment.

v) **Clarification of the conditions for privatising R&D organisations.** A simplistic vision of the market economy has pushed certain groups, including some within government, to advocate massive privatisation of research structures, while others, such as laboratory directors or local authorities, in search of the most immediately available resources, actively engage in the sale or rental of their facilities. Clear rules must be established to determine what is an inalienable public good, even though private interests may be granted use of it (notably basic research infrastructure) and what can be privatised (notably industrial R&D facilities that might be integrated into firms).

vi) **Rebuilding the social sciences.** These have been profoundly affected by the current crisis and require particular state support. A parallel special effort is needed to elaborate structures and programmes for training in management and related skills needed in the new economic context.

3. Promoting and protecting innovation

Innovation is the act of creating a new technical product and diffusing it through the economy, and, in the long term, it is the principal source of wealth. The underlying idea may arise from a scientific finding, from experience acquired in production, or from a perceived need. In any case, innovation is never the simple application of a research result even when it derives from one. It requires design, market studies, and engineering, and an entrepreneur to bring the project to fruition. In the market economy painfully being put in place, conditions for innovation are poor and have as yet received little attention from the authorities. It is likely that the evolution of the whole system of production is involved, including the issues of privatisation, employment, foreign investment, etc. However, the questions touching more directly on entrepreneurship and technology would deserve closer attention. We will restrict ourselves to recommendations in three areas:

i) **Multiplication of local programmes for furnishing innovators with the information and material support they most urgently need.** In most cases, they need information on markets (including foreign ones), the conditions applying to exports and imports, standards, and sources of materials and components. In practical terms, they need, in the first instance, premises equipped with telephone and international fax, located if possible near their commercial, scientific, and technical associates. This does not imply large-scale projects, such as vast technology parks, but small-scale assistance programmes (which might find their place in "business centres"), for which the local and regional authorities should have the primary responsibility, in terms both of conception and of realisation.

ii) **Improvement of all the legal and judicial sources of protection for inventors and entrepreneurs.** The law on patents, passed in autumn 1992, is a good law, but if no enforcement mechanisms are instituted (control, means of recourse, penalties), it will be of little use. Protection for entrepreneurs through control mechanisms and means of recourse is also urgently needed, as part of the framework of anti-monopoly laws that are in the course of application or elaboration. These laws must also be able to be used to put pressure on suppliers, clients, and subcontractors, who still often are, in the great majority of cases, state enterprises or organisations. **It is in fact a true judicial system – applicable to science, technology and industry – that must be constructed today in Russia.** Administrative and bureaucratic obstacles – commercial licences, travel visas, export and import licences, etc. – that continue to hinder the activity of "average" entrepreneurs should also be reduced.

iii) **Promotion of product quality in all possible ways.** This is essential to inducing change in the practice of producers and consumers and to creating progressively a domestic market that corresponds to international norms. The massive promotion of quality means the enactment of appropriate regulations, the setting up of systematic controls and penalties, sensitisation to quality control in factories, etc. To a certain extent, the infrastructures exist, in the form of regional agencies of the State Committee on Standards, to which this task could usefully be given.

4. **Developing and modernising civil technology**

In the former economic and political system, civil technology did not receive the attention it deserved. Given the current economic crisis and the lack of resources, it would seem best to stick to essentials and to what can bring in the most at the least cost. From this viewpoint, it is suggested that, in the immediate future, a few key actions should be taken:

i) **Reducing significantly the barriers to information and to access to scientific and technical centres formerly engaged in military work.** Much information and many inventions and technologies have no reason to remain secret for strategic reasons and can usefully be exploited for civil purposes. The opening up of military science and technology, which has already begun, is doubtless the best means to facilitate the conversion of the institutes concerned and to serve the needs of society at the same time.

ii) **Actively developing public procurement policies in some sectors** which can have a non-negligible influence on the morale of the population and, at the same time, develop a sense of quality: health, housing, transport, and energy (energy saving in particular) offer various possibilities that should, however, be chosen with care, given the lack of resources. These are sectors in which the State and public authorities are, in principle, able to intervene.

iii) **Modernising the telecommunications infrastructures and services,** as the basis for an entrepreneurial economy and the precondition for active foreign investment. This implies defining coherent plans, realistic regulations and standards, and creating conditions for the effective mobilisation of foreign capital and techniques.

5. **Internationalising Russian science and technology**

Broad and rapid internationalisation of Russian science and technology is, in effect, the factor most certain to ensure the survival of its finest elements. This is also true for other countries with advanced science and technology, but it gains greater relief in the light of the present circumstances in Russia. Not only do the funds engaged by the industrialised countries – by both the private and the public sectors – constitute an important source of financing, but foreign involvement also helps select the Russian institutes, teams, and firms capable of being integrated at a high level in the international community. In other words, the internationalisation of Russian S&T introduces an essential "reality principle" to the process of transformation. On the other hand, the opening of Russian S&T often has a beneficial effect on the concerned foreign communities. Internationalisation will occur all the more easily and efficiently if a number of measures – to be elaborated and put in place in collaboration with foreign partners – are drawn up.

i) **Facilitating the transfer of foreign funds for research and development,** which are presently subject to very high taxes. This situation creates considerable obstacles to collaboration and leads to fraudulent behaviour on both sides. Taxation on these transfers should eventually be completely removed, and conditions that ensure the greatest possible transparency should be created. As has been seen, there are already many public and private bilateral and multilateral initiatives. For the most part, they are directed to individual researchers or to specific projects with teams selected by the foreign funding bodies. The Russian authorities are unhappy with this state of affairs, as they consider that international co-operation does not facilitate the restructuring process. One may thus wonder whether, as a complement to existing initiatives, a fund should not be set up for conveying certain international financing, to be managed by representatives of OECD countries and the appropriate organisms in Russia and used to realise strategic restructuring operations in the more general framework of the Government's scientific policy.

ii) **Systematically setting up arrangement that help avoid brain drain in collaborative operations** of all kinds (joint projects, twinning of laboratories, invitations to Russian teachers, etc.) and facilitate the return of Russian researchers to their home country. Their return should also be helped on the Russian side by protective measures (for example, maintaining their positions). International research activities and courses or seminars could also usefully be organised in Russia in certain fields. In general, it would seem to be of benefit to all that scientific and technical competencies are maintained in Russia.

iii) **Furnishing broad international support to doctoral candidates and young professors in the universities** in order to ensure the renewal and replacement of the scientific work force and to encourage the development of research capacities in the university.

iv) **Providing the support necessary to maintain the science infrastructure** (libraries, large-scale facilities, computer networks, databases) **and developing material support of technical research,** which is considerably hampered by the lack of technical components and materials that are generally unavailable in Russia. Moreover, a special effort should be made by concerned international organisations to help Russia establish norms and certification procedures so that they can gradually integrate global markets.

v) **Aiding by adequate legal arrangements, and perhaps by financial incentives, the subcontracting proposed by private industry in the OECD countries, both in research and in production.** For research, such subcontracting is advantageous for both parties; private foreign industry can benefit from remarkable competencies at reasonable cost, while Russian science is confronted with the demands of ''western'' industry. In production, subcontracting facilitates the diffusion of standards and methods in force in OECD countries, and it gives the firms involved access to the vast Russian domestic market. The development of this scientific and industrial subcontracting will require the establishment of a balanced and rigorous regime for intellectual and industrial property, as well as adequate conditions for foreign investment. Jointly developed data banks that collect information on Russian R&D and technology (institutes, teams, coverage, results achieved, etc.) would also be useful.

vi) **Seeking ''co-operative equilibria'' in a certain number of key sectors that carry considerable weight in world industry and trade,** and in which Russians have, for various reasons, a comparative advantage over OECD countries, but which they are unable to exploit without the support of these countries. For example, Russia has competitive facilities in certain sectors, but costs are unknown and undervalued; other activities are conducted more cheaply because with little regard for environmental considerations. The disequilibria that will eventually arise in world markets are ruinous for the OECD countries, which will naturally be inclined to reconsider their aid to the Russians if co-operative approaches, advantageous for both parties, are not established.

vii) **Finally, for the OECD countries, it is important, in the implementation of technical assistance, to help bring about a change of views by carefully targeting actions on well-defined sites** and carrying them out with the necessary determination and continuity. In this respect, it is advisable to work out very concrete operations (conversion of firms, transformation of institutes, opening of training centres) that can bring about changes in perception and stimulate emulation throughout the country. It would be good, as well, if the different international organisations involved in various ways would better co-ordinate their efforts on specific sites, following agreement among the parties concerned. In addition, the collection and exchange of appropriate information, on the part of both Russia and concerned international organisations, would facilitate the identification of interesting sites and stimulate informal co-ordination.

6. Improving statistics and accounting practices

It is not possible to guide either a macroeconomic or a microeconomic system without statistics and clear accounting practices that measure inputs and outputs.

i) **In the area of national science and technology indicators,** significant progress has been made in attempting to reach OECD standards (the Frascati Manual). These efforts should be pursued, and more precise information should be obtained for the defence sector, so that the data are comparable to what is available for the other large technological countries. Measures of output (inventions, innovations, establishment of firms, etc.) should also be developed, beginning with specific sectors and sites in order to refine the methodology. Support for indicator work both from Russia and from foreign organisations would be most useful.

ii) It is essential that **the presentation of the state budget for science and technology** contains as clear and detailed a description as possible of financial allocations so that the priorities are brought out. It is also important that budgeted or promised funds are actually delivered and used for the purposes specified. The same effort at transparency is necessary in the *ex post* presentation of the use of funds, in particular in Parliament, in accordance with procedures used in OECD countries.

iii) **Finally, it is indispendable to develop full accounting practices in laboratories and firms in order to measure total real costs (including indirect subsidies) of the scientific and technical endeavour.** It is essential for good management of financial resources in a situation of serious economic recession; it is also crucial for establishing realistic pricing practices, and in particular for forestalling tendencies to "dumping" on international markets.

*

* *

Many observers are likely to agree to many of the proposals formulated above; on more than one issue, these proposals also undoubtedly reflect the projects and the philosophy of the Russian authorities. Yet, as has been remarked – in all countries, including those apparently the most open to reform – **the most difficult is not so much to determine what is to be done but rather to discover how to do it,** given the extent to which the changes clash with opinions, habits, and acquired advantages. Yet is not change the very source of life?

Annex 1

Tables

Table 1. **R&D expenditure in Russia in 1990 and 1991**
In current roubles (millions)

	1990	1991
Current R&D expenditure	10 902.9	18 197.7
Capital R&D expenditure	2 174.8	1 793.0
Total R&D expenditure	13 077.7	19 990.7
As a per cent of gross domestic product (GDP)	2.1	1.54

Source: Centre for Science Research and Statistics, Moscow (1993) for Background Report, *Science, Technology and Innovation Policies: Federation of Russia* (data established in line with OECD standards).

Table 2. **Distribution of R&D personnel by USSR republics: 1980, 1985, 1990**
(per cent)

Republics	Personnel employed in "Science and scientific services"			1990		Number of R&D specialists per 1 000 workers of republics
	1980	1985	1990	Employees of R&D organisations	R&D specialists*	
USSR	100.0	100.0	100.0	100.0	100.0	14
Russia	71.0	70.4	70.8	68.4	68.2	17
Other republics	29.3	28.8	31.2	31.4	n.a.	
Ukraine	13.7	15.4	13.7	17.4	17.4	13
Baltic republics	2.4	2.1	2.4	n.a.	n.a.	n.a.
Asian republics	6.8	6.3	6.4	4.5	4.4	n.a.
Caucasus republics	3.8	3.9	3.9	3.3	3.6	n.a.

* Employed in research activities; excludes administrative and technical personnel.
n.a.: Not available.
Sources: *Research and Development in the USSR. Data Book: 1990*, Centre for Science Research and Statistics, 1992, pp. 13, 17; *Science in Russia Today and Tomorrow*; Analytical Centre, Russian Academy of Sciences, 1992.

Table 3. **Geographic distribution of Russia's R&D expenditures and personnel by economic regions, 1991**

Geographic area	Expenditures		Researchers [1]	
	Millions of current roubles	% [2]	Thousands	% [2]
Russia, total	20 105.9	100.0	878.5	100.0
North	270.5	1.3	10.5	1.2
North-West	3 087.5	15.4	147.1	16.8
St Petersburg	3 011.7	15.0	142.0	16.2
Central	8 883.2	44.2	377.2	42.9
Moscow	6 118.9	30.4	263.2	30.0
Volga-Viatka	797.0	4.0	36.1	4.1
Central Black Earth	537.2	2.7	28.3	3.2
Volga	1 478.2	7.3	69.4	7.9
North Caucasus	818.7	4.1	41.5	4.7
Urals	1 600.5	8.0	66.3	7.5
West Siberian	1 645.7	8.2	60.1	6.8
East Siberian	451.3	2.2	21.3	2.4
Far East	484.4	2.4	18.5	2.1
Kaliningrad	51.7	0.3	2.1	0.2

1. Excluding teaching staff working in higher education as part-time researchers.
2. May not sum to 100 because of rounding.
Source: *Science and Technology in Russia: 1991*, Centre for Science Research and Statistics, 1992, Tables 3.9 and 4.18, and Figure 31.

Table 4. **Researchers employed by different types of performing organisations, 1990-91**

	Number of researchers,* thousands		Distribution of researchers,* per cent	
	Total	Doctors and candidates of sciences	Total	Doctors and candidates of sciences
1990				
Russia, total	992.6	142.5	100.0	100.0
Industrial R&D organisations	745.4	76.7	75.1	53.8
Industrial enterprises	68.6	2.5	6.9	1.7
Academies of Sciences	107.5	45.0	10.8	31.6
Higher education	71.1	18.3	7.2	12.9
1991				
Russia, total	878.5	134.2	100.0	100.0
Industrial R&D organisations	649.7	69.6	74.0	51.9
Industrial enterprises	59.0	1.6	6.7	1.2
Academies of Sciences	109.0	46.5	12.4	34.7
Higher education	60.8	16.5	6.9	12.2
1991/1990, %				
Russia, total	88.5	94.2	–	–
Industrial R&D organisations	87.2	90.7	–	–
Industrial enterprises	86.0	64.0	–	–
Academies of Sciences	101.4	103.3	–	–
Higher education	85.5	90.2	–	–

* R&D specialists employed in research activities, excluding administrative and technical personnel.
Source: *Science and Technology in Russia: 1991*, Centre for Science Research and Statistics, 1992, Table 3.3.

Table 5. **Financing of R&D projects by type of activity and performer, 1990 and 1991**
(per cent)

	Industrial R&D organisations	Industrial enterprises	Academies of Sciences	Higher education	R&D total
			1990		
R&D, total	78.5	4.9	10.5	6.1	100.0
Basic research	23.8	0.9	62.4	12.9	100.0
Applied research	75.8	2.6	10.6	10.9	100.0
Development	88.7	6.9	2.4	2.0	100.0
			1991		
R&D, total	78.4	4.2	11.8	5.5	100.0
Basic research	18.1	0.3	64.6	17.0	100.0
Applied research	74.9	1.7	14.1	9.3	100.0
Development	90.3	6.3	1.9	1.5	100.0

Source: Science and Technology in Russia: 1991, Centre for Science Research and Statistics, 1992, Table 4.10.

Table 6. **Approximate number of institutes in the Russian R&D system**
(1992)

Academies of Sciences	586
(*of which* Russian Academy of Sciences)	(350)
Higher education	450
Branch research institute (civil and defence)	3 128
(*of which* R&D divisions of industrial firms)	(400)

Sources: Background Report, OECD S&T Policy Review; Centre for Science Research and Statistics and Analytical Centre, Russian Academy of Sciences.

Table 7. **Average monthly salaries in the sector "Science and scientific services"**

	1970	1980	1985	1988	1989	1990	1991	1992
Average monthly salary in the sector "Science and scientific services", roubles	143.2	184.9	209.9	256.6	314.3	351.9	558.0	4 108.5
As a per cent of salary:								
in the national economy (= 100)	113.6	104.1	104.2	109.1	121.5	118.6	105.3	70.9
in industry (= 100)	105.3	96.7	96.3	102.5	114.2	113.2	96.2	59.9
in construction (= 100)	92.7	86.8	84.8	83.7	92.9	93.6	82.3	51.7

Source: Centre for Science Research and Statistics.

Table 8. **Payments for and receipts from trade in licences**

(data for the former USSR, million roubles*)

	1980	1985	1990
Hard currency receipts from exports of licences	14.5	20.1	48.8
Total currency receipts from exports of licences	79.8	150.7	211.2
Currency payments for imports of licences	132	127	205.8

* Estimated in roubles using the official exchange rates of the USSR State Bank.
Source: *Research and Development in the USSR, Data Book 1990,* Centre for Science Research and Statistics, 1992, p. 51.

Table 9. **Estimated budgets of bilateral collaboration programmes in science and technology of selected OECD countries with Russia**

	1992 (million dollars)	1993 (million dollars)
France[1]	22	22
Germany[2]	45	66
Japan[3]	7	36
USA[4]	54	228
Canada[5]	25	25
UK[6]	2	2
Finland[7]	6	6
Total[8]	161	385

The estimates are converted into US$ (conversion rates for 1992 as listed in the *Economic Outlook* of June 1993 and as at June 1993). These estimates are indicative only, given the different reporting procedures and differences in coverage.
1. France does not include nuclear or space collaborations but includes the university sector: figures were not provided for 1993; 1993 figure is an estimate.
2. Germany does not include the university sector but does include the nuclear sector.
3. Japan includes the nuclear sector and related safety operations.
4. The United States does not include operations by NASA and space collaborations under its responsibility; nuclear safety is not included.
5. Out of C$ 150 million its 1992-96 programme of C$ 150 million over five years.
6. UK provided figures for 1993 only: 1992 an estimate.
7. Figures provided by the Finnish Centre for Technology (VTT).
8. These figures do not include contributions to multilateral programmes, such as ITER, ISTC.
Source: OECD Secretariat.

Annex 2

Summary of the discussion of the evaluation report
Moscow, 21-22 September 1993

This summary, drafted by the OECD Secretariat, presents the main points raised in the meeting held in Moscow to discuss the Evaluation Report. The meeting was attended by some 60 representatives from OECD Member countries (see the List of Participants at the end of this annex) and a Russian delegation led by the Minister of Science and Technological Policy, Mr. Saltykov, who co-chaired the meeting with Professor J.-L. Lions of the Collège de France, President of the International Mathematical Union.

In the course of the discussion, participants were able to express their views on the report and to add complementary information. Also brought out were points of agreement and disagreement between assessments by the Russian authorities and the OECD countries. After some introductory observations on the objectives of the study and how it was conducted, the discussion followed the general organisation of the successive chapters of the report.

Introductory observations

Minister Saltykov emphasised that in addressing the current problems of science and technology in Russia, the long history of their development must be kept in mind. Russia has a long-standing scientific tradition which goes back at least to the 17th century. Then, during the Communist era, enormous investments were made in this domain, considered as one of the primary factors in the economic development of the country and its power on the world stage. In the difficult transition the country is now undergoing, the public authorities are concerned to preserve the best part of this immense legacy and to adapt science and technology to the conditions of a market economy. The Minister thanked the OECD for this study, which is proving to be very useful for the elaboration and implementation of reforms.

Mr. Boright, Deputy Assistant Secretary of State of the United States and Vice-President of the OECD Committee for Science and Technological Policy, confirmed the importance that the OECD and its Member countries attached to this work, which has been accomplished in record time and to which considerable resources have been dedicated. In emphasising the respect inspired by Russian science and the modesty appropriate to efforts to evaluate it, he spoke for the international community and the experts who carried out the study.

Mr. Tanaka, Director for Science, Technology, and Industry at the OECD, thanked the Russian authorities who had shown great initiative and daring in asking the OECD to undertake this study. As always, the OECD tried not to avoid the difficulties to be faced as Russia attempts to save and reorient its science and technology, but to facilitate debate and discussion among all the concerned communities. He insisted also on the need to avoid simplistic ideologies, since, if the market economy is indispensable to a nation's survival, it may not suffice to restart its development.

Mr. Shorin, Chairman of the Parliamentary Commission on Science and Education, underscored that his presence at the meeting was proof of the Parliament's active support of the reform policies of the Ministry of Science and Technological Policy. He also indicated that the OECD report was to be used in the Parliamentary discussions planned for October 1993 on science policy and, in particular, on the Academy of Sciences.

Mr. Laverov, Vice-President of the Academy of Sciences, indicated that a quite clear view of how to guide the reorganisation of the Russian R&D system had now emerged and that it had deteriorated far less than some people in Russia and elsewhere were saying.

Mr. Piskunov, Vice-Chairman of the State Committee for Industrial Policy and Scientific Leader of the Analytical Centre of the Academy of Sciences, who collaborated in the preparation of the Background Report to the study, stressed that the OECD recommendations had been prepared in close collaboration with the Russian experts and, to a significant degree, also reflected their views.

Mr. Mindeli, Director of the Centre for Science Research and Statistics and also involved in the study, emphasised that the study had usefully stimulated the harmonisation of Russia's analytical and accounting practices with those of the OECD.

Professor Lions, who guided the discussion, requested that, in the debate on the various sections of the report, the speakers focus on the relevant recommendations, as they appear in the final chapter.

General problems and institutional aspects

The participants agreed that Russian science and technology was suffering greatly in the transition period and even appeared, to some extent, as one of the very first victims of the transformation process and related reforms.

However, the problems were all the more serious because of the sector's excessive size, as *Professor Cooper,* who contributed to the report, emphasised in introducing it. The S&T sector's reduction, along with the dimensions appropriate to a stabilised economy in line with international standards was, without doubt, the most discussed issue of the meeting. The data presented in the Evaluation Report, in particular in Box A, evoked a number of remarks from the Russian delegation.

First, a wider range of definitions and computations should be used when evaluating categories such as science or culture. It is not possible to plan Russian expenditures by comparison with those of the United States. The national characteristics of the country, its history, its geographic situation, and the relation between scientific capacity, broadly understood, and industrial and military capacity should be taken into consideration. If all these factors are taken into account, a reduction of the magnitude apparently suggested by the Evaluation Report is impossible.

Second, a large reduction in Russian science and technology seems inevitable, but it should be carried out only after measures have been taken for evaluating the nation's scientific property. Before that, national S&T policy should be strictly oriented to preserving and maintaining the accumulated scientific capability. The first task is to establish priorities and determine what the vitally important elements of the system are, not only in order to develop them, but primarily so that the ineffective parts of the scientific system can be reduced appropriately. This should be the main goal of state S&T policy, in order to oppose the present "darwinistic" approach.

Third, when dealing with the reduction of Russian science, two criteria in particular should be applied. First, the annual outflow of scientific personnel should not exceed 20 per cent; and second, a time limit should be set that takes account of the interval needed to elaborate and put into practice the initial stages of selection.

In response to these remarks, the Secretariat stated that in furnishing these indications, it had simply wished to give some information that might enlighten the various actors. It had in no way intended to make predictions or give recommendations. It was certainly appropriate to take account of the specific characteristics of Russia, notably the relatively labour-intensive nature of its economy in general and of its research system in particular. This tradition would perhaps allow Russia to maintain a body of scientists and technicians somewhat larger than the real size of its economy would seem to support. The latter was furthermore difficult to estimate and to compare with that of other nations.

It was also agreed that present economic and political conditions made it extremely difficult to implement an efficient adjustment policy for the science and technology sector independently of other economic adjustment policies. They also considerably limited the actions available to the public authorities responsible for this area. From this perspective, the need for co-ordination with the Government and the authorities in charge of other spheres of activity was discussed at length. In fact, as the Japanese delegation, in particular, pointed out, there is a great deal of interdependence between the reforms and the policies carried out. Budgetary and financial policy, strategic doctrine, privatisation of the economy, among others, will have an important influence on the future of science and technology.

Minister Saltykov indicated that, in his opinion, the commission formed several months ago, which he chaired and which brought together some 60 representatives of the most relevant ministries, had proved rather cumbersome and was not in a position to make decisions. This is why the co-ordinating mechanism of an interministerial body at the level of the Prime Minister, proposed by the Evaluation Report, seemed to him a judicious suggestion. However, certain delegations, particularly the German delegation, expressed serious doubts about the effectiveness and the value of such a mechanism. Instead, they favoured a structure under which the Ministry of Science and Technological Policy has primary competence and plays a leading role with respect to the other concerned ministries. It takes responsibility (in the Cabinet and before Parliament) for the general principles governing publicly financed areas of R&D.

The German delegation also insisted on the need to take sufficient account of the federal structure of Russia. Thus, the report's recommendations concerning the regions should not concentrate so much on the establishment and the capacities of regional agencies and authorities but rather on the need for a highly decentralised research system safeguarded by the Constitution. A system that divides the competencies of the state in the field of R&D promotion between the Federation and the regions would provide only limited room for centralised decision making; it would be funded by a variety of sources; and intermediate bodies would be needed to take responsibility for the co-ordination of research policy matters across the Federation. The Secretariat replied that in view of the present uncertain constitutional circumstances, it had been difficult to be precise, especially when information was unavailable on the financial resources effectively to be allocated to the regions in the new distribution of power between the centre and the periphery.

The problem of defining priorities and carrying out the corresponding evaluations was raised by the Russian delegation. To their idea of establishing a parliamentary evaluation office similar to the US Office of Technology Assessment (OTA), the American delegation replied that the office in question had substantial resources and that it did not participate directly in the formulation of choices, even if it played an important advisory role for Congress. On the question of the evaluation of scientific and technological strengths and weaknesses, it was obvious that a great country like Russia, which, like the United States, had covered all scientific domains in depth, would find it very difficult to select some at the expense of others. However, if intermediate evaluations were conducted of sectors, institutes, and projects, using rigorous methods, it should be possible gradually to make the necessary choices.

The science base and the research structures

Several delegations underscored the need for a better integration of the research system and higher education and expressed concern for the renewal and the training of high-level scientists in the present crisis situation. *Professor Balzer,* who contributed to the Evaluation Report, proposed that massive aid be given by the international community with a view to creating perhaps some 10 000 positions for teaching staff and researchers, half to be recruited in the Academy of Sciences and half in the university system. He also proposed that the amount of the budget accorded to the Russian Foundation for Fundamental Science should be raised from the current 3 per cent of the civil research budget to 10 per cent.

Some delegations voiced concern over the future of the branch institutes. They emphasised the need for co-ordinated action to identify and support strategic S&T institutions by the government bodies concerned with S&T, industry policy and privatisation, and they raised questions about the appropriateness of the programme of federal research centres launched by the ministry. *Minister Saltykov* indicated that there had been something of an emergency situation, where it had been necessary to take immediate measures to save important national and even international scientific assets. About 30 institutes had finally been singled out as an initial measure; the Ministry supports 40 per cent of the personnel costs of the institutes, which must find complementary funding through contracts with Russian or foreign organisations.

Professor Marchuk, former president of the Academy of Sciences of the USSR, indicated that certain large Academy institutes with large installations were contemplating joining the programme and requesting affiliation as federal research centres. He also mentioned the formation of an Association of the Academies of Sciences of the Republics of the former Soviet Union that renewed in a formal manner the ties of the past. He also appreciated the use of the word "shrine" to describe the Academy in the English version of the report. Some, including Russian participants, regretted the religious connotation and would have preferred a term closer to the French "haut lieu".

The fate of the social sciences in Russia drew the attention of several delegations, and they wished to see a specific point made about this issue in the recommendations.

Finally, the question of brain drain was an important subject of debate. The analysts from the Centre for Science Research and Statistics argued that the figure of 30 000 temporary and permanent emigrants, indicated in Box C, was far too high. In general, it was agreed that the qualitative aspects should attract more concern that the quantitative aspects of the issue. Further, worry about external brain drain (abroad) was accompanied by concerns over internal brain drain (within the Russian economy itself). *Minister Saltykov* indicated that, however painful the process might be, it seemed to him to fall under the law of the conservation of matter and energy. The scientists who left the research sector furnished other sectors of the burgeoning market economy (firms, banks, etc.) with qualified managerial personnel who were needed and could not be found elsewhere. It was not possible both to reduce the R&D sector substantially and keep its labour force, even if some of the best scientists left. In addition, it was a good sign that these people found new employment without too much difficulty.

As for the problem of the long-term renewal of the nation's science base, the minister mentioned surveys indicating that 10 to 15 per cent of the student population were planning to study scientific disciplines. This proportion was not very different from that of the 1980s, although it was also the case that the overall student population itself had diminished by 10 to 15 per cent since then.

The innovation climate and technological development

Mr. Linnakko, a contributor to the report, stressed that it presented several important traits of the climate in which the spirit of innovation and enterprise manifested itself at present. It quite intentionally did not analyse in detail an innovation system which was being completely transformed. *Professor Marchuk* offered some figures on the past working of the system. It was estimated that overall, in the planned economy, 28 per cent of the investments went to innovation, with significant variations in various sectors: 40 to 50 per cent in defence and between 15 and 20 per cent in civil activities. Renewal of equipment and products varied from five years for branches such as electronics to 15 years for steel-making.

The establishment of a climate favourable to innovation in the present economic and political situation in Russia elicited many questions from participants. They asked, in particular, how significant industrial research could be instituted within the firms themselves. Some participants emphasised that this was conditional on more general industrial policy measures and that the current orientations of this policy were very unclear. The role of small businesses was brought up, and it was noted that although they have proliferated, many are sheltered in institutes and by bureaucracy and vegetate to some extent.

Minister Saltykov also stressed that hyperinflation impeded investment in research and innovation. He agreed that not enough had yet been done to establish the legal framework necessary for protecting and supporting entrepreneurs and innovators. These services, related to patents, standards, exports, etc., were furnished by foreign consultants or firms at exorbitant prices, and he called on the international community and multilateral bodies to give greater help in this area. Some participants wondered, however, whether it would not take Russia a long time to set up such legal and institutional frameworks, given the absence of a tradition of a society founded on civil rights.

The importance of establishing a system of efficient telecommunications was also emphasised by several delegations, who supported the views contained in the Evaluation Report on this topic.

Finally, several delegations from OECD countries that dedicate a significant part of their R&D effort to defence brought out the painful difficulties their own countries are encountering as they convert military to civil activities. They questioned Russia's ability to bring about conversion and voiced doubts about the potential applications in certain sectors mentioned in the report (environment, health, etc.). In general, it was noted that the report had treated these questions in a relatively superficial manner, and the wish to see certain sectors treated in greater depth in a possible follow-up to the study was expressed.

Internationalisation and international co-operation

Questions of international co-operation occupied a separate session. *Ms. Prerost,* a consultant to the Secretariat, presented the results of a survey of science and technology co-operation in major OECD countries with Russia.

The Russian authorities, through *Mr. Yakobashvili,* Vice-Minister for International Scientific Co-operation, expressed their doubts about the estimates of financial flows for scientific co-operation from the OECD countries which were contained in Box B and which they found inflated. They also wished foreign operators to use central organisational channels, which were better able, in their opinion, to co-ordinate and to choose the best Russian partners. They also indicated their plan to establish control over the exporting of technology from Russia very soon.

Several delegations from the OECD Member countries reiterated their desire for improved procedures for transferring funds and expressed their concerns over excessive control on co-operation by the central authorities. They even wondered whether the centralising and bureaucratic tendencies of the past had really disappeared. In addition, several delegations regretted that the legal and regulatory obstacles to the establishment and full operation of the International for Science and Technology Center had still not been removed.

Professor Rabkin, consultant to the Secretariat, who collected, within the framework of the study, about 40 position papers from representatives of the Russian scientific community, underscored the importance that these people attached to the support of the international community. They saw in this support the means to escape the control of the central authorities. Also stressed was the importance of minimising the obstacles to research and innovation activities of foreign firms, which were seen to play a fundamental role in the present situation.

In the Russian delegation, voices were raised to express disagreement with certain practices engaged in by foreign businesses which completely take over industrial property rights when they fund and employ Russian teams in Russia. Disappointment was also expressed about the fact that some assistance proposed to the Russians is rather humiliating, given the modest nature of the subsidies proposed.

It was nonetheless agreed that, despite the divergences between representatives of the OECD countries and the Russian authorities, the study and the related investigations had made it possible to initiate a dialogue that would best continue if transparency was ensured by all participants.

Conclusion

Professor Lions drew the conclusions of the discussions and noted that, overall, participants had generally approved the report, including its recommendations. However, some of the latter needed to be reconsidered or supplemented in the light of the discussion. It was also clear that the situation was changing rapidly and that its evolutionary character should be clearly acknowledged in the report.

List of participants

Chair

Mr. Boris Saltykov, Minister for Science and Technological Policy, Federation of Russia

Mr. Jacques-Louis Lions, Professor, Collège de France

Russian Delegation

Mr. Andrei Fonotov, First Vice-Minister for Science and Technological Policy

Mr. Nicolai Laverov, Vice-President, Academy of Sciences

Mr. Guri Marchuk, Vice-President, Academy of Sciences

Mr. Levan Mindeli, Director, Centre for Science Research and Statistics

Mr. Dimitri Piskunov, Vice-Chairman, State Committee for Industrial Policy

Mr. Vladimir Shorin, Chairman, Commission for Science and Education, Supreme Soviet

Mr. Zarub A. Yakobashvili, Vice-Minister for Science and Technological Policy

Mr. Boris Yurlov, Head of Department, Ministry of Science and Technological Policy

*

* *

Mr. Leonid Gokhberg, Centre for Science Research and Statistics

Mr. Vladimir Kisselev, Consultant, Ministry of Science and Technological Policy

Ms. Victoria Kisseleva, Analytical Centre of the Russian Academy of Sciences

Mr. Igor Nikolaev, Head of Division, Ministry of Science and Technological Policy

Foreign Delegates

Austria
Dr. Peter Bachmaier

Belgium
Dr. Monnik Desmeth

Canada
Mr. Rick Mann
Mr. Alexander Leonidovich Petrov

Denmark
Mr. Knud Larsen
Mr. Thomas Bisgaard

Finland
Mr. Alpo Kuparinen
Ms. Elisabeth Helander
Mr. Markku Linna
Mr. Antero Inkari
Mr. Esko-Olavi Seppälä

France
Ms. Martine Beurton
Mr. Edouard Brezin
Mr. Pierre Baruch
Ms. Elisabeth Legrand
Mr. Georges Prokhoroff
Mr. Massenet
Ms. Brigitte Godelier

Germany
Dr. Jürgen Arnold
Dr. Heinrich Vogel
Dr. Uwe Meyer

Japan
Mr. Iwao Ohashi
Mr. Masaki Komurasaki
Mr. Shigeharu Kato
Mr. Shinichi Kuroki

The Netherlands
Dr. Noé Van Hulst
Dr. Ingrid Dillo
Dr. Egbert Oldenboom

Norway
Mr. Leif Egil Westgaard
Ms. Karen Nossum Bie
Mr. Jan-Eilert Askeröi
Mr. Runar Jensen

Switzerland
Prof. Verena Meyer

Uinted Kingdom
Dr. Paul Potter
Dr. Joan Hare

United States
Mr. John Phillips Boright
Ms. Christine Glenday
Mr. William Kingkade
Mr. Jeff Schweitzer
Mr. Vlad Sambaiew
Ms. Carolyn Shettle
Mr. Gary Waxmonsky
Ms. Dorothy Zinberg

Commission of the European Communities

Mr. Luigi Massimo

Delegates from non-member countries

Hungary
Ms. Veronika Csillagné Bálint

Slovak Republic
Dr. Rudolf Demovic
Dr. Dusan Valachovic

Poland
Dr. Jan Kozlowski

International Organisations
Dr. Alain Jubier
NATO

Mr. Vladislav Kotchetkov
UNESCO

Mr. Charles Blitzer
World Bank

Mr. Glenn Schweitzer
Mr. Alain Gerard
Mr. Atsushi Shaku
ICST

Mr. Jean-Michel Chasseriaux
INTAS[1]

Dr. Il'dar Karimov
IIASA[2]

OECD Consultants
Mr. Harley Balzer
Professor Julian Cooper
Mr. Ilkka Linnakko

Ms. Sandra Prerost
Prof. Yakov Rabkin
Mr. Christian Sautter

OECD Secretariat
    Mr. Nobuo Tanaka, Director, Science, Technology and Industry
    Mr. Michael W. Oborne, Head, Science and Technology Policy Division
    Mr. Daniel Malkin, Head, Economic Analysis and Statistics Division
    Mr. Jean-Eric Aubert, Co-ordinator of the Study, Science and Technology Policy Division
    Mr. Martin Salamon, Administrator, Science, Technology and Industry Directorate; Centre for Co-operation with Economies in Transition

---

1. International Association for the Promotion of Cooperation with Scientists from the Independent States of the Former Soviet Union.
2. International Institute for Applied Systems Analysis.

MAIN SALES OUTLETS OF OECD PUBLICATIONS
PRINCIPAUX POINTS DE VENTE DES PUBLICATIONS DE L'OCDE

ARGENTINA – ARGENTINE
Carlos Hirsch S.R.L.
Galería Güemes, Florida 165, 4° Piso
1333 Buenos Aires   Tel. (1) 331.1787 y 331.2391
Telefax: (1) 331.1787

AUSTRALIA – AUSTRALIE
D.A. Information Services
648 Whitehorse Road, P.O.B 163
Mitcham, Victoria 3132   Tel. (03) 873.4411
Telefax: (03) 873.5679

AUSTRIA – AUTRICHE
Gerold & Co.
Graben 31
Wien I   Tel. (0222) 533.50.14

BELGIUM – BELGIQUE
Jean De Lannoy
Avenue du Roi 202
B-1060 Bruxelles   Tel. (02) 538.51.69/538.08.41
Telefax: (02) 538.08.41

CANADA
Renouf Publishing Company Ltd.
1294 Algoma Road
Ottawa, ON K1B 3W8   Tel. (613) 741.4333
Telefax: (613) 741.5439
Stores:
61 Sparks Street
Ottawa, ON K1P 5R1   Tel. (613) 238.8985
211 Yonge Street
Toronto, ON M5B 1M4   Tel. (416) 363.3171
Telefax: (416)363.59.63
Les Éditions La Liberté Inc.
3020 Chemin Sainte-Foy
Sainte-Foy, PQ G1X 3V6   Tel. (418) 658.3763
Telefax: (418) 658.3763

Federal Publications Inc.
165 University Avenue, Suite 701
Toronto, ON M5H 3B8   Tel. (416) 860.1611
Telefax: (416) 860.1608

Les Publications Fédérales
1185 Université
Montréal, QC H3B 3A7   Tel. (514) 954.1633
Telefax : (514) 954.1635

CHINA – CHINE
China National Publications Import
Export Corporation (CNPIEC)
16 Gongti E. Road, Chaoyang District
P.O. Box 88 or 50
Beijing 100704 PR   Tel. (01) 506.6688
Telefax: (01) 506.3101

DENMARK – DANEMARK
Munksgaard Book and Subscription Service
35, Nørre Søgade, P.O. Box 2148
DK-1016 København K   Tel. (33) 12.85.70
Telefax: (33) 12.93.87

FINLAND – FINLANDE
Akateeminen Kirjakauppa
Keskuskatu 1, P.O. Box 128
00100 Helsinki

Subscription Services/Agence d'abonnements :
P.O. Box 23
00371 Helsinki   Tel. (358 0) 12141
Telefax: (358 0) 121.4450

FRANCE
OECD/OCDE
Mail Orders/Commandes par correspondance:
2, rue André-Pascal
75775 Paris Cedex 16   Tel. (33-1) 45.24.82.00
Telefax: (33-1) 45.24.81.76   or (33-1) 45.24.85.00
Telex: 640048 OCDE

OECD Bookshop/Librairie de l'OCDE :
33, rue Octave-Feuillet
75016 Paris   Tel. (33-1) 45.24.81.67
(33-1) 45.24.81.81

Documentation Française
29, quai Voltaire
75007 Paris   Tel. 40.15.70.00

Gibert Jeune (Droit-Économie)
6, place Saint-Michel
75006 Paris   Tel. 43.25.91.19

Librairie du Commerce International
10, avenue d'Iéna
75016 Paris   Tel. 40.73.34.60

Librairie Dunod
Université Paris-Dauphine
Place du Maréchal de Lattre de Tassigny
75016 Paris   Tel. (1) 44.05.40.13

Librairie Lavoisier
11, rue Lavoisier
75008 Paris   Tel. 42.65.39.95

Librairie L.G.D.J. - Montchrestien
20, rue Soufflot
75005 Paris   Tel. 46.33.89.85

Librairie des Sciences Politiques
30, rue Saint-Guillaume
75007 Paris   Tel. 45.48.36.02

P.U.F.
49, boulevard Saint-Michel
75005 Paris   Tel. 43.25.83.40

Librairie de l'Université
12a, rue Nazareth
13100 Aix-en-Provence   Tel. (16) 42.26.18.08

Documentation Française
165, rue Garibaldi
69003 Lyon   Tel. (16) 78.63.32.23

Librairie Decitre
29, place Bellecour
69002 Lyon   Tel. (16) 72.40.54.54

GERMANY – ALLEMAGNE
OECD Publications and Information Centre
August-Bebel-Allee 6
D-53175 Bonn 2   Tel. (0228) 959.120
Telefax: (0228) 959.12.17

GREECE – GRÈCE
Librairie Kauffmann
Mavrokordatou 9
106 78 Athens   Tel. (01) 32.55.321
Telefax: (01) 36.33.967

HONG-KONG
Swindon Book Co. Ltd.
13–15 Lock Road
Kowloon, Hong Kong   Tel. 366.80.31
Telefax: 739.49.75

HUNGARY – HONGRIE
Euro Info Service
POB 1271
1464 Budapest   Tel. (1) 111.62.16
Telefax : (1) 111.60.61

ICELAND – ISLANDE
Mál Mog Menning
Laugavegi 18, Pósthólf 392
121 Reykjavik   Tel. 162.35.23

INDIA – INDE
Oxford Book and Stationery Co.
Scindia House
New Delhi 110001   Tel.(11) 331.5896/5308
Telefax: (11) 332.5993

17 Park Street
Calcutta 700016   Tel. 240832

INDONESIA – INDONÉSIE
Pdii-Lipi
P.O. Box 269/JKSMG/88
Jakarta 12790   Tel. 583467
Telex: 62 875

IRELAND – IRLANDE
TDC Publishers – Library Suppliers
12 North Frederick Street
Dublin 1   Tel. (01) 874.48.35
Telefax: (01) 874.84.16

ISRAEL
Electronic Publications only
Publications électroniques seulement
Sophist Systems Ltd.
71 Allenby Street
Tel-Aviv 65134   Tel. 3-29.00.21
Telefax: 3-29.92.39

ITALY – ITALIE
Libreria Commissionaria Sansoni
Via Duca di Calabria 1/1
50125 Firenze   Tel. (055) 64.54.15
Telefax: (055) 64.12.57
Via Bartolini 29
20155 Milano   Tel. (02) 36.50.83

Editrice e Libreria Herder
Piazza Montecitorio 120
00186 Roma   Tel. 679.46.28
Telefax: 678.47.51

Libreria Hoepli
Via Hoepli 5
20121 Milano   Tel. (02) 86.54.46
Telefax: (02) 805.28.86

Libreria Scientifica
Dott. Lucio de Biasio 'Aeiou'
Via Coronelli, 6
20146 Milano   Tel. (02) 48.95.45.52
Telefax: (02) 48.95.45.48

JAPAN – JAPON
OECD Publications and Information Centre
Landic Akasaka Building
2-3-4 Akasaka, Minato-ku
Tokyo 107   Tel. (81.3) 3586.2016
Telefax: (81.3) 3584.7929

KOREA – CORÉE
Kyobo Book Centre Co. Ltd.
P.O. Box 1658, Kwang Hwa Moon
Seoul   Tel. 730.78.91
Telefax: 735.00.30

MALAYSIA – MALAISIE
Co-operative Bookshop Ltd.
University of Malaya
P.O. Box 1127, Jalan Pantai Baru
59700 Kuala Lumpur
Malaysia   Tel. 756.5000/756.5425
Telefax: 757.3661

MEXICO – MEXIQUE
Revistas y Periodicos Internacionales S.A. de C.V.
Florencia 57 - 1004
Mexico, D.F. 06600   Tel. 207.81.00
Telefax : 208.39.79

NETHERLANDS – PAYS-BAS
SDU Uitgeverij Plantijnstraat
Externe Fondsen
Postbus 20014
2500 EA's-Gravenhage   Tel. (070) 37.89.880
Voor bestellingen: Telefax: (070) 34.75.778

**NEW ZEALAND
NOUVELLE-ZÉLANDE**
Legislation Services
P.O. Box 12418
Thorndon, Wellington	Tel. (04) 496.5652
	Telefax: (04) 496.5698

NORWAY – NORVÈGE
Narvesen Info Center – NIC
Bertrand Narvesens vei 2
P.O. Box 6125 Etterstad
0602 Oslo 6	Tel. (022) 57.33.00
	Telefax: (022) 68.19.01

PAKISTAN
Mirza Book Agency
65 Shahrah Quaid-E-Azam
Lahore 54000	Tel. (42) 353.601
	Telefax: (42) 231.730

PHILIPPINE – PHILIPPINES
International Book Center
5th Floor, Filipinas Life Bldg.
Ayala Avenue
Metro Manila	Tel. 81.96.76
	Telex 23312 RHP PH

PORTUGAL
Livraria Portugal
Rua do Carmo 70-74
Apart. 2681
1200 Lisboa	Tel.: (01) 347.49.82/5
	Telefax: (01) 347.02.64

SINGAPORE – SINGAPOUR
Gower Asia Pacific Pte Ltd.
Golden Wheel Building
41, Kallang Pudding Road, No. 04-03
Singapore 1334	Tel. 741.5166
	Telefax: 742.9356

SPAIN – ESPAGNE
Mundi-Prensa Libros S.A.
Castelló 37, Apartado 1223
Madrid 28001	Tel. (91) 431.33.99
	Telefax: (91) 575.39.98

Libreria Internacional AEDOS
Consejo de Ciento 391
08009 – Barcelona	Tel. (93) 488.30.09
	Telefax: (93) 487.76.59

Llibreria de la Generalitat
Palau Moja
Rambla dels Estudis, 118
08002 – Barcelona
	(Subscripcions) Tel. (93) 318.80.12
	(Publicacions) Tel. (93) 302.67.23
	Telefax: (93) 412.18.54

SRI LANKA
Centre for Policy Research
c/o Colombo Agencies Ltd.
No. 300-304, Galle Road
Colombo 3	Tel. (1) 574240, 573551-2
	Telefax: (1) 575394, 510711

SWEDEN – SUÈDE
Fritzes Information Center
Box 16356
Regeringsgatan 12
106 47 Stockholm	Tel. (08) 690.90.90
	Telefax: (08) 20.50.21
Subscription Agency/Agence d'abonnements :
Wennergren-Williams Info AB
P.O. Box 1305
171 25 Solna	Tel. (08) 705.97.50
	Téléfax : (08) 27.00.71

SWITZERLAND – SUISSE
Maditec S.A. (Books and Periodicals - Livres
et périodiques)
Chemin des Palettes 4
Case postale 266
1020 Renens	Tel. (021) 635.08.65
	Telefax: (021) 635.07.80

Librairie Payot S.A.
4, place Pépinet
CP 3212
1002 Lausanne	Tel. (021) 341.33.48
	Telefax: (021) 341.33.45

Librairie Unilivres
6, rue de Candolle
1205 Genève	Tel. (022) 320.26.23
	Telefax: (022) 329.73.18

Subscription Agency/Agence d'abonnements :
Dynapresse Marketing S.A.
38 avenue Vibert
1227 Carouge	Tel.: (022) 308.07.89
	Telefax : (022) 308.07.99

See also – Voir aussi :
OECD Publications and Information Centre
August-Bebel-Allee 6
D-53175 Bonn 2 (Germany)	Tel. (0228) 959.120
	Telefax: (0228) 959.12.17

TAIWAN – FORMOSE
Good Faith Worldwide Int'l. Co. Ltd.
9th Floor, No. 118, Sec. 2
Chung Hsiao E. Road
Taipei	Tel. (02) 391.7396/391.7397
	Telefax: (02) 394.9176

THAILAND – THAÏLANDE
Suksit Siam Co. Ltd.
113, 115 Fuang Nakhon Rd.
Opp. Wat Rajbopith
Bangkok 10200	Tel. (662) 225.9531/2
	Telefax: (662) 222.5188

TURKEY – TURQUIE
Kültür Yayinlari Is-Türk Ltd. Sti.
Atatürk Bulvari No. 191/Kat 13
Kavaklidere/Ankara	Tel. 428.11.40 Ext. 2458
Dolmabahce Cad. No. 29
Besiktas/Istanbul	Tel. 260.71.88
	Telex: 43482B

UNITED KINGDOM – ROYAUME-UNI
HMSO
Gen. enquiries	Tel. (071) 873 0011
Postal orders only:
P.O. Box 276, London SW8 5DT
Personal Callers HMSO Bookshop
49 High Holborn, London WC1V 6HB
	Telefax: (071) 873 8200
Branches at: Belfast, Birmingham, Bristol, Edinburgh, Manchester

UNITED STATES – ÉTATS-UNIS
OECD Publications and Information Centre
2001 L Street N.W., Suite 700
Washington, D.C. 20036-4910 Tel. (202) 785.6323
	Telefax: (202) 785.0350

VENEZUELA
Libreria del Este
Avda F. Miranda 52, Aptdo. 60337
Edificio Galipán
Caracas 106	Tel. 951.1705/951.2307/951.1297
	Telegram: Libreste Caracas

Subscription to OECD periodicals may also be placed through main subscription agencies.

Les abonnements aux publications périodiques de l'OCDE peuvent être souscrits auprès des principales agences d'abonnement.

Orders and inquiries from countries where Distributors have not yet been appointed should be sent to: OECD Publications Service, 2 rue André-Pascal, 75775 Paris Cedex 16, France.

Les commandes provenant de pays où l'OCDE n'a pas encore désigné de distributeur devraient être adressées à : OCDE, Service des Publications, 2, rue André-Pascal, 75775 Paris Cedex 16, France.

2-1994

OECD PUBLICATIONS, 2 rue André-Pascal, 75775 PARIS CEDEX 16
PRINTED IN FRANCE
(14 94 02 1) ISBN 92-64-14081-6 - No. 46967 1994